Mathematical Essays

By Kermit Rose

Third Edition

ISBN-13:
978-1500735388

ISBN-10:
1500735388

Self Published on Create Space

Table of Contents

Introduction to the Irrational Plane

Complex plane and irrational numbers.

Where to begin?

Begin in the middle, and work backwards and forwards.

What are complex numbers?

Complex numbers are the sum of real numbers and imaginary numbers.

What are imaginary numbers?
What are real numbers?

Examples of natural numbers are 0, 1, 2, 3, ...
These are also called whole numbers or integers.

What are rational numbers?

Rational numbers are the result of dividing an integer by an integer.

1/2, 3/4, 204312/1221, ... are rational numbers.

What (again) are real numbers?

Imagine an infinitely long line. The line you draw on paper is an illustration of this infinitely long line in your imagination.

Pick a point on this infinitely long line, and stipulate that it is the middle of the line. (Any point will do since the line is infinitely long.)

Place the number 0 on that point.

Now pick another point, (to the right of 0), and place the number 1 on that point. The distance between 0 and 1, is called the unit distance, and that distance is also called 1.

Move 1 unit to the right of the point 1, and place the number 2 on the line. Move one more unit to the right, and place the number 3.

Continue forever.

Imagine that now we have placed all the positive integers on the line.
The next step is to place -1, -2, -3, ... on the line in a way similar to how we placed the positive integers.

Alternatively, you can imagine that we have initiated an automatic machine to place the integers, and it is working while we have gone on to do other things. This machine has already placed more integers than we are capable of imagining within a lifetime.

So now that any integer we can imagine has been placed on the line, we can think about the next step.

The process of division comes naturally from the process of multiplication.

3 * 4 = 12
leads naturally to 12/4 = 3 and 12/3 = 4.

What about 13/5 and 7/2?

These are numbers between integers.

13/5 is somewhere between 10/5 = 2, and 15/5 = 3.
7/2 is somewhere between 6/2 = 3, and 8/2 = 4.

Now we construct another machine to busily start placing all the rational numbers on the line.

We very cleverly figure out instructions for the machine to do all the simple, easy to imagine rational numbers first.

It starts out with 1/2, -1/2, 1/3, -1/3, 2/3, -2/3, 1/4, -1/4, 3/4, -3/4, 1/5, -1/5, 2/5, -2/5, 3/5, -3/5, 4/5, -4/5, etc.

Ok, now all the rational numbers that we can imagine have been placed on the number line. And the remaining rational numbers are still being placed by the machine.

Have all the points of the line been numbered?

No.

There is a point on the line between 1 and 2 that corresponds to the square root of 2.

The square root of 2 is not equal to any integer divided by another integer.

The proof is simple.

A number between 1 and 2 is not an integer.
Consider any number between 1 and 2 that is one integer divided by another.

Lets give it a name. Say $q = p/r$, where p and r are integers, and r does not divide p.

then $q*q = (p/r) * (p/r) = (p * p)/(r*r)$ is also not an integer, because if r does not divide p, neither can (r*r) divide (p*p).

Since q*q is not an integer, it cannot be equal to 2.

So, there are points on the line that have not been assigned a rational number or integer number. Incidentally, integers are also examples of rational numbers. $2 = 2/1$, $3 = 3/1$, etc.

The numbers we assign to those points that are not rational numbers, are the irrational numbers.

There are infinitely many irrational numbers, although we, of course, can in our lifetime, name only finitely many of them.

Other examples of irrational numbers are
sqrt(3), sqrt(5), sqrt(7), sqrt(11), ... pi, e, (e is the base of natural logarithms).

Now that the irrational real numbers have been described, and the rational real numbers have been described,

we can say,

a real number is a rational real number or an irrational real number.

A real number is a number that can be placed on the number line.

What (again) are imaginary numbers?

A few hundred years ago, someone worked out the quadratic formula for solving general quadratic equations, such as,

$x^2 + 5x + 6 = 0$.

$x^2 - 6x + 8 = 0$.

$A x^2 + B x + C = 0$,
where A,B,C have not yet been specified,
and we wish to find x, in terms of A, B, and C.

This general formula sometimes gave solutions that required that we take the square root of a negative number.

The square of a negative number is positive.
The square of a positive number is positive.

There does not exist any real number whose square is negative.

These numbers, whose square is negative, are not less than 0, are not greater than 0. Hence they were called imaginary.

Today, we still call them "imaginary", but credit them with as much reality as we credit the "real" numbers.

The sum of a real number and an imaginary number is called a complex number.

How are the imaginary numbers graphed?

Since they are neither less than or greater than zero, we place them on a vertical line, perpendicular to our number line. The imaginary axis line goes through 0.

Thus we have a two dimensional graph of the complex numbers.
This two dimensional graph is called a plane.

The irrational plane consists of all the points of the graphed plane that corresponded to points of the form

irrational real + imaginary number

or

real number + irrational imaginary number.

Mathematical induction

The basic idea of mathematical induction is this:
Suppose you wish to prove that a statement is true of all positive integers.
If you can prove that it is true for the integer k=1, and you can also prove that, in general, whenever it is true of the integer k, then it is also true of the integer (k+1), then you have proven it for all positive integers.
Let's look at this in detail.
You have proven the statement for the integer k=1.
Since whenever the statement is true of the integer k, then it is also true of the integer (k+1), we know that if the statement is true of k=1, it is also true of k=2.
Therefore, because of that general step, we make a chain.
Because the statement is true of k=2, it is true of k=3.
Because the statement is true of k=3, it is true of k=4.
Because the statement is true of k=4, it is true of k=5.
Because the statement is true of k=5, it is true of k=6.
Etc.

In this way, you automatically ascend to every finite positive integer.
That is you have proven that the statement is true for all positive integers.

This is why mathematical induction works. We have established the principle of mathematical induction.

If you prove the statement is true of the integer 1,and you also prove, in general, that whenever the statement is true of the integer k, it is also true of the integer (k+1), then you have therefore proven that it is true of all positive integers.

One example frequently given is to prove that for all positive integers n, that $1 + 2 + 3 + 4 + ... + n = n*(n+1)/2$

To implement the mathematical induction algorithm for this example, we first prove the formula is true for k = 1.

Then we prove that whenever it is true for some integer k, that $1 + 2 + 3 + ... + k = k * (k+1)/2$, then it is also true that
$1 + 2 + 3 + ... + k + (k+1) = (k+1) * ((k+1)+1)/2$.
How is this different from what we were asked to prove? How is this different than assuming true what we were asked to prove?

We were asked to prove that the formula is true for all positive integers.

We presume that it is true for only one unspecified, positive integer.

Then we are to use that presumption that it is true for some unspecified integer k, to prove that it is also true for the integer (k+1).

In this example, we proceed as follows.

We are to prove that, for the specific integer k=1, that the sum of the integers from 1 to k is equal to k(k+1)/2, and we are to prove that if for some unspecified integer k,
$$1 + 2 + 3 + ... + k = k * (k+1)/2$$
then
$$1 + 2 + 3 + ... + k + (k+1) = (k+1)*(k+2)/2.$$

It's easy to prove this formula for the integer k = 1.
The sum of the integers from 1 to 1 is just 1.

Since 1 = 1 *2/2, the formula is true for k = 1.

To prove the general case, we replace the
$$1 + 2 + 3 + ... + k \quad \text{in}$$
$$1 + 2 + 3 + ... + k + (k+1 \quad \text{by}$$
$$k * (k+1)/2$$

which gives us the equation
$$1 + 2 + 3 + ... + k + (k+1)$$
$$= k*(k+1)/2 + (k+1)$$

By factoring out the (k+1) we get
$$1 + 2 + 3 + ... + k + (k+1)$$
$$= k * (k+1)/2 + (k+1)$$
$$= k*(1/2) * (k+1) + (k+1)$$
$$= (k * 1/2 + 1) * (k+1)$$
Note that we can simplify
$$(k * 1/2 + 1)$$
to
$$(k/2 + 1) = (k+2)/2$$
That is, we now have
$$1 + 2 + 3 + ... + k + (k+1)$$

= k * (k+1)/2 + (k+1)
= (k * 1/2 + 1) * (k+1)
= ((k+2)/2) * (k+1)

Change the order of multiplication to get
((k+2)/2)* (k+1) = (k+1)*(k+2)/2

Thus , we have proven that

1 + 2 + 3 + ... + k + (k+1)
= k * (k+1)/2 + (k+1)
= (k * 1/2 + 1) * (k+1)
= ((k+2)/2) * (k+1)
= (k+1)*(k+2)/2

We have proven that the formula is valid for (k+1).
That is, we have proven that whenever the formula is valid for one unspecified integer k, then it is also true of (k+1).

This completes the mathematical induction because we can follow the chain
true of 1 ==> true of 2 ==> true of 3 ==> true of 4 ==> .....

Generalization of Quadratic Equation

You have probably seen the derivation of the quadratic formula solution to

$A x^2 + B x + C = 0$.

Divide through by A

$x^2 + (B/A) x + (C/A) = 0$

Subtract (C/A) from both sides of the equation.

$x^2 + (B/A) x = - (C/A)$

Complete the Square by adding $(B/2A)^2$

$x^2 + (B/A) x + (B/2A)^2 = (B/2A)^2 - (C/A)$

$(x (B/2A))^2 = (B/2A)^2 - (C/A)^2$

Take the square root of the left side.

$(x + (B/2A)) = +$ or $-$ sqrt$((B/2A)^2 - (C/A))$

Now consider a generalization of the quadratic equation.
$A x y + B x + C y = D$

We go through almost the identical steps to solve it.
Divide through by A.
$x y + (B/A) x + (C/A) y = (D/A)$

Complete the product by adding $(B C/A^{2)}$
$x y + (B/A) x + (C/A) y + (B C / A^2) = (B C / A^2) + (D / A)$

Factor the left side.
$(x + (C/A)) (y + (B/A)) = (B C / A^2) + (D / A)$

Multiply through by the denominator A^2
$(A x + C)(A y + B) = (B C + A D)$

To find x and y, factor (BC+AD) = p q
and set
$(A x + C) = p$
and $(A y + B) = q$.

$x = (p-C)/A$ and $y = (q - B)/A$.

A Special Number Sequence

Pick any odd positive integer.

It is the difference of two squares.

For example, 55 = 8*8 - 3*3

Then we can make a number series by adding up and down as follows:

8 * 8 - 3 * 3 = 55

7 * 9 - 2 * 4 = 55

6 * 10 - 1 * 5 = 55

5 * 11 - 0 * 6 = 55

4 * 12 + 1 * 7 = 55

3 * 13 + 2 * 8 = 55

2 * 14 + 3 * 9 = 55

1 * 15 + 4 * 10 = 55

0 * 16 + 5 * 11 = 55

In General,
if $t^2 - s^2 = z$,
for any integer r,
$(t^2 - r^2) - (s^2 - r^2) = t^2 - s^2 = z$
$(t - r)(t + r) - (s - r)(s + r) = z$

Eight Queens Problem

Put eight queens on the chessboard so that no two queens may attack each other.

This is equivalent to making a permutation of the digits 1, 2, 3, 4, 5, 6, 7, 8 such that no two digits have a position difference equal to their arithmetic difference.

For example, in the sequence 1, 3, 5, 3 is adjacent to 1. Their position difference is 1. But (3-1) = 2, is not 1. Similarly (5-3)= 2 is not 1.

The sequence 1, 3, 4 would not be permitted because 3 and 4 are adjacent. Their position difference is 1. 4-3 = 1. 1,3,4 is a sequence in which the position difference of 3 and 4 is the same as their arithmetic difference.

No two digits that have positions differ by 2, have themselves, a difference of 2
The sequence 3, 1, 5 would not be permitted because the 3 and 5 have a difference in position of 2, and 5-3 = 2.

No two digits that have position differ by 3, have themselves, a difference of 3.
The sequence 1, 7, 2, 8, 4 would not be permitted because 7 and 4 have positions that differ by 3, and 7-4 = 3.

No two digits that have position differ by 4, have themselves, a difference of 4.
The sequence 2 8 1 7 6 would not be permitted because 2 and 6 have positions that differ by 4, and 6-2 = 4.

No two digits that have position differ by 5, have themselves, a difference of 5.
The sequence 3,5,2,7,1,8 would not be permitted because 3 and 8 have positions that differ by 5, and 8-3 = 5.

No two digits that have position differ by 6, have themselves, a difference of 6.
The sequence 2, 5, 3, 1, 4,6, 8 would not be permitted because 2 and 8 have positions that differ by 6 and 8-2 = 6.

No two digits that have position differ by 7, have themselves, a difference of 7.

The sequence 1, 7, 4, 6, 3, 5, 2, 8 would not be permitted because 1 and 8 have positions that differ by 7, and 8-1 = 7.

There are 92 solutions.

Four of the 92 solutions are symmetric with respect to rotation and reflections of the chessboard.

One of the symmetric solutions is 35281746.

Put the first queen in row 3 of column 1.

Put the second queen in row 5 of column 2.

Put the third queen in row 2 of column 3.

Put the fourth queen in row 8 of column 4.

Put the fifth queen in row 1 of column 5.

Put the sixth queen in row 7 of column 6.

Put the seventh queen in row 4 of column 7.

Put the eighth queen in row 6 of column 8.

Rotating the board clockwise 90 degrees,

will result in seeing the solution,

46827135.

That is, new column 1 has a queen in row 4;

new column 2 has a queen in row 6,

new column 3 has a queen in row 1, etc.

Rotating the board clockwise again 90 degrees, results in the solution

35281746, the same as the original, since this is the symmetric solution.

Rotating the board clockwise again 90 degrees, results in the solution,
46827135, which is the same as the rotation of the original by 90 degrees.

To get the other two relatives of the symmetric solution, reflect from either the front of the
chessboard or the back of the chessboard.

This is equivalent to subtracting each digit from 9.

The Four relatives of the symmetric solution are:

35281746
64718253
53172864
46827135

The other 88 solutions are related to each other in similar ways.

The 88 solutions fall into 11 sets of 8 solutions related by rotation and reflections.

When given this puzzle decades ago, I first found the symmetric solution. Later I systematically found the other 88.

My systematic method consisted of scanning the permutations of 12345678, in numerical order, and skipping over the permutations that do not corresponds to solutions.

The smallest permutation found to be solution is:
15863724.

The 8 relatives for this permutation are:
15863724
84136275 by subtracting each digit from 9.
57263148 by reversing digits.
42736851 by subtracting each digit from 9.
82417536 by transposing, equivalent to rotation and reflection.
17582463 by subtracting each digit from 9.
36428571 by reversing digits.
63571428 by subtracting each digit from 9.

Transposing of 42736851 means make the new solution such that:
The 1 is in position 4,
the 2 is in position 2,
the 3 is in position 7,
the 4 is in position 3,
the 5 is in position 6,
the 6 is in position 8,
the 7 is in position 5,
the 8 is in position 1,
yielding 82417536.

We have found 4+8=12 of the 92 solutions.

Transform 15863724 by moving the 4 from the end to the beginning.
This yields 41586372.
This gives 8 more solutions as
41586372
58413627 by subtracting each digit from 9.
72631485 by reversing digits.
27368514 by subtracting each digit from 9.
71386425 by transposing, = rotation and reflection.
28613574 by subtracting each digit from 9.
47531682 by reversing digits.
52468317 by subtracting each digit from 9.

We have found 12+8 = 20 of the 92 solutions.

Transform 82417536 by moving the 6 from the end to the beginning.

This yields 68241753.

Note that this permutation is not any of the 8 associated with the 41586372, the first result of the end round transformation.

From 68241753 we have 8 associations.
68241753
31758246 by subtracting each digit from 9.
64285713 by reversing digits.
35714286 by subtracting each digit from 9.
53847162 by transposing.
46152837 by subtracting each digit from 9.
73825164 by reversing digits.
26174835 by subtracting each digit from 9.

We have found 20+8 = 28 of the 92 solutions.

Transform the 68241753 by moving the 3 from the end to the beginning.

This yields 36824175.

Note that this permutation is not any that were previously found.

36824175
63175824 by subtracting each digit from 9.
42857136 by reversing digits.
57142863 by subtracting each digit from 9.
35841726 by transposing.
64158273 by subtracting each digit from 9.
37285146 by reversing digits.
62714853 by subtracting each digit from 9.

We have found 28+8 = 36 of the 92 solutions.
Transform 64158273 by moving the 6 from the beginning to the end, yielding
41582736.

41582736
58417263 by subtracting each digit from 9.
36271485 by reversing digits.
63728514 by subtracting each digit from 9.
74286135 by transposing.
25713864 by subtracting each digit from 9.

46831752 by reversing digits.
53168247 by subtracting each digit from 9.

We have found 44 of the 92 solutions.

Transform 74286135 by moving the 7 from beginning to end,
yielding 42861357.

42861357
57138642 by subtracting each digit from 9.
24683175 by reversing digits.
75316824 by subtracting each digit from 9.
47382516 by transposing.
52617483 by subtracting each digit from 9.
38471625 by reversing digits.
61528374 by subtracting each digit from 9.

We have found 52 of the 92 solutions.

We seem to have exhausted the finding of new solutions by moving a digit from one end of the
permutation to the other.

Transform the previously found solution 15863724
to 51863724 by reversing the first two digits.

It is a solution.

51863724
48136275 by subtracting each digit from 9.
57263184 by reversing digits.
42736815 by subtracting each digit from 9.
72418536 by transposing.
27581463 by subtracting each digit from 9.
36418572 by reversing digits.
63581427 by subtracting each digit from 9.

We have found 60 of the 92 solutions.

Originally I found the 92 solutions in one afternoon by using the chessboard to scan a subset of
the 8!=40320 permutations to check if they were solutions. I used the backtracking shortcut to
skip over permutations which had subsections that were not solutions.

The web page

http://en.wikipedia.org/wiki/Eight_queens_puzzle

gives all the solutions and discusses additional short cuts to finding the solutions.

Representatives of the 12 sets of solutions,
11 sets of 8, and 1 set of 4, are

1	24683175
2	17468253
3	17582463
4	41582736
5	51842736
6	31758246
7	51468273
8	71386425
9	51863724
10	57142863
11	63184275
12	53172864

Note that the last one shown is the symmetric solution, recognized because the reversal of digits is the same as subtracting each digit from 9.

Meaning of Equality

What does it mean for two different number expressions to be equal?

It means that the two different number expressions are each reducible to the same number.

What does it mean that $2 + 5 = 3 + 4$?

$2 + 5$ reduces to $2 + (4 + 1) = (2 + 4) + 1 = 6 + 1 = 7$

$3 + 4$ reduces to $3 + (3 + 1) = (3 + 3) + 1 = 6 + 1 = 7$

There are many degrees of equality.

If we say that two things are identical, we mean that they are the same in all intrinsic characteristics. They would still be different by being in different places or different times.

In a weaker use of "identical", we refer to identical twins that have the same genetic makeup, but different environmental experiences.

Another way in which two things could be equal is if they are different names for the same thing. Some writers use this point of view when they say that "2 + 5" and "3 + 4" are different names for the number 7.

Another example of the "different names for the same thing" is if I say things like the following.
"Let p be a variable name for a number.
At the beginning of an algorithm let p take on the value of 7."
OR
"Let A be a constant equal to 7.
At the beginning of the algorithm, set $p = A$."

When we introduce remainder arithmetic, we see yet another level of equality.

9 and 2 have the same remainder when divided by 7.

$9 = 2$ modulo 7.

All concepts of equality, (worthy of the name), have certain characteristics in common.

Consider the set of objects to which this particular concept of equality is being applied.
It might be the set of integers. It might be a larger set of numbers. It might be other objects that we usually do not call numbers.
The thing that all concepts of equality have in common is that the equality relation partitions the set to which the equality relation is applied.
All elements within the same partition are considered the same with respect to everything that matters.

For example, with the "have the same remainder, when divided by 7" relation, the integers, {0,7,-7,14,-14,21,-21,...} are the same with respect to "remainder when divided by 7".

All the integers {2,9,16,23,...} are the same with respect to "remainder when divided by 7".

In a social setting, when one states that men and women have equal rights, we really mean the following. Your gender is not to be considered important with respect to determining what your rights are to be.

We also may extend the concept of equality to games.
Here are two games that I will call equal.

game 1: Tic Tac Toe.

A pair of vertical lines is crossed perpendicularly by a pair of horizontal lines, producing 9 squares.

Players take turns placing either an "O" or an "X" into one of the squares. Player who makes "three in a row", a straight line, of three of their symbols, wins the game.

game 2: fifteen.

Blocks or pieces of paper, numbered 1 to 9, are placed between the two players. Players take turns selecting a number from the pieces.
Once a piece is taken, it is no longer available for the other player to take.
The first player to be able to have three pieces add to 15, wins the game.

These two games are equal.
Here is the proof of why I can say that these two games are equal.
Consider the 3 by 3 square of numbers
8 1 6
3 5 7
4 9 2

It is the 9 numbers placed on the tic tac toe board.
As can easily be seen, every straight line of numbers that would have won a tic tac toe game has the numbers add up to 15.
And, it is also true the other way around.
Every set of three numbers on this tic tac toe board,
that add to 15, lie on a straight line.

So now you know that every winning tactic for "tic tac toe" can be translated into a winning tactic for the game "fifteen".

Rule of 72?

The rule of 72 is: The number of compounding periods required to double your principle is 72% divided by the interest rate. This rule is accurate only if the interest rate is near 7.85%.

This chart shows which number to divide by the interest rate to calculate the number of periods to reach a given multiple of the principle.

Desired multiple -->	2	3	4	5
Interest Rate				
0.10%	69.35%	109.92%	138.70%	161.02%
0.20%	69.38%	109.97%	138.77%	161.10%
0.30%	69.42%	110.03%	138.84%	161.19%
0.40%	69.45%	110.08%	138.91%	161.27%
0.50%	69.49%	110.14%	138.98%	161.35%
0.60%	69.52%	110.19%	139.04%	161.43%
0.70%	69.56%	110.25%	139.11%	161.51%
0.80%	69.59%	110.30%	139.18%	161.59%
0.90%	69.63%	110.35%	139.25%	161.67%
1.00%	69.66%	110.41%	139.32%	161.75%
1.10%	69.70%	110.46%	139.39%	161.83%
1.20%	69.73%	110.52%	139.46%	161.91%
1.30%	69.76%	110.57%	139.53%	161.99%
1.40%	69.80%	110.63%	139.60%	162.07%
1.50%	69.83%	110.68%	139.67%	162.15%
1.60%	69.87%	110.74%	139.74%	162.23%
1.70%	69.90%	110.79%	139.80%	162.31%
1.80%	69.94%	110.85%	139.87%	162.39%
1.90%	69.97%	110.90%	139.94%	162.47%
2.00%	70.01%	110.96%	140.01%	162.55%
2.10%	70.04%	111.01%	140.08%	162.63%
2.20%	70.07%	111.07%	140.15%	162.71%
2.30%	70.11%	111.12%	140.22%	162.79%
2.40%	70.14%	111.17%	140.29%	162.87%
2.50%	70.18%	111.23%	140.36%	162.95%
2.60%	70.21%	111.28%	140.42%	163.03%
2.70%	70.25%	111.34%	140.49%	163.11%
2.80%	70.28%	111.39%	140.56%	163.19%
2.90%	70.31%	111.45%	140.63%	163.27%
3.00%	70.35%	111.50%	140.70%	163.35%
3.10%	70.38%	111.56%	140.77%	163.43%
3.20%	70.42%	111.61%	140.84%	163.51%
3.30%	70.45%	111.66%	140.90%	163.58%

3.40%	70.49%	111.72%	140.97%	163.66%
3.50%	70.52%	111.77%	141.04%	163.74%
3.60%	70.56%	111.83%	141.11%	163.82%
3.70%	70.59%	111.88%	141.18%	163.90%
3.80%	70.62%	111.94%	141.25%	163.98%
3.90%	70.66%	111.99%	141.32%	164.06%
4.00%	70.69%	112.04%	141.38%	164.14%
4.10%	70.73%	112.10%	141.45%	164.22%
4.20%	70.76%	112.15%	141.52%	164.30%
4.30%	70.79%	112.21%	141.59%	164.38%
4.40%	70.83%	112.26%	141.66%	164.46%
4.50%	70.86%	112.31%	141.73%	164.54%
4.60%	70.90%	112.37%	141.79%	164.62%
4.70%	70.93%	112.42%	141.86%	164.70%
4.80%	70.97%	112.48%	141.93%	164.78%
4.90%	71.00%	112.53%	142.00%	164.86%
5.00%	71.03%	112.59%	142.07%	164.93%
5.10%	71.07%	112.64%	142.14%	165.01%
5.20%	71.10%	112.69%	142.20%	165.09%
5.30%	71.14%	112.75%	142.27%	165.17%
5.40%	71.17%	112.80%	142.34%	165.25%
5.50%	71.20%	112.86%	142.41%	165.33%
5.60%	71.24%	112.91%	142.48%	165.41%
5.70%	71.27%	112.96%	142.54%	165.49%
5.80%	71.31%	113.02%	142.61%	165.57%
5.90%	71.34%	113.07%	142.68%	165.65%
6.00%	71.37%	113.13%	142.75%	165.73%
6.10%	71.41%	113.18%	142.82%	165.80%
6.20%	71.44%	113.23%	142.88%	165.88%
6.30%	71.48%	113.29%	142.95%	165.96%
6.40%	71.51%	113.34%	143.02%	166.04%
6.50%	71.54%	113.39%	143.09%	166.12%
6.60%	71.58%	113.45%	143.16%	166.20%
6.70%	71.61%	113.50%	143.22%	166.28%
6.80%	71.65%	113.56%	143.29%	166.36%
6.90%	71.68%	113.61%	143.36%	166.43%
7.00%	71.71%	113.66%	143.43%	166.51%
7.10%	71.75%	113.72%	143.49%	166.59%
7.20%	71.78%	113.77%	143.56%	166.67%
7.30%	71.81%	113.82%	143.63%	166.75%
7.40%	71.85%	113.88%	143.70%	166.83%
7.50%	71.88%	113.93%	143.77%	166.91%
7.60%	71.92%	113.98%	143.83%	166.98%
7.70%	71.95%	114.04%	143.90%	167.06%
7.80%	71.98%	114.09%	143.97%	167.14%
7.85%	72.00%	114.12%	144.00%	167.18%

7.90%	72.02%	114.15%	144.04%	167.22%
8.00%	72.05%	114.20%	144.10%	167.30%
8.10%	72.09%	114.25%	144.17%	167.38%
8.20%	72.12%	114.31%	144.24%	167.46%
8.30%	72.15%	114.36%	144.31%	167.53%
8.40%	72.19%	114.41%	144.37%	167.61%
8.50%	72.22%	114.47%	144.44%	167.69%
8.60%	72.25%	114.52%	144.51%	167.77%
8.70%	72.29%	114.57%	144.58%	167.85%
8.80%	72.32%	114.63%	144.64%	167.93%
8.90%	72.36%	114.68%	144.71%	168.00%
9.00%	72.39%	114.73%	144.78%	168.08%
9.10%	72.42%	114.79%	144.85%	168.16%
9.20%	72.46%	114.84%	144.91%	168.24%
9.30%	72.49%	114.89%	144.98%	168.32%
9.40%	72.52%	114.95%	145.05%	168.39%
9.50%	72.56%	115.00%	145.11%	168.47%
9.60%	72.59%	115.05%	145.18%	168.55%
9.70%	72.62%	115.11%	145.25%	168.63%
9.80%	72.66%	115.16%	145.32%	168.71%
9.90%	72.69%	115.21%	145.38%	168.79%
10.00%	72.73%	115.27%	145.45%	168.86%
10.10%	72.76%	115.32%	145.52%	168.94%
10.20%	72.79%	115.37%	145.59%	169.02%
10.30%	72.83%	115.43%	145.65%	169.10%
10.40%	72.86%	115.48%	145.72%	169.17%
10.50%	72.89%	115.53%	145.79%	169.25%
10.60%	72.93%	115.59%	145.85%	169.33%
10.70%	72.96%	115.64%	145.92%	169.41%
10.80%	72.99%	115.69%	145.99%	169.49%
10.90%	73.03%	115.75%	146.05%	169.56%
11.00%	73.06%	115.80%	146.12%	169.64%
11.10%	73.09%	115.85%	146.19%	169.72%
11.20%	73.13%	115.90%	146.26%	169.80%
11.30%	73.16%	115.96%	146.32%	169.87%
11.40%	73.19%	116.01%	146.39%	169.95%
11.50%	73.23%	116.06%	146.46%	170.03%
11.60%	73.26%	116.12%	146.52%	170.11%
11.70%	73.29%	116.17%	146.59%	170.19%
11.80%	73.33%	116.22%	146.66%	170.26%
11.90%	73.36%	116.28%	146.72%	170.34%
12.00%	73.40%	116.33%	146.79%	170.42%
12.10%	73.43%	116.38%	146.86%	170.50%
12.20%	73.46%	116.43%	146.92%	170.57%
12.30%	73.50%	116.49%	146.99%	170.65%
12.40%	73.53%	116.54%	147.06%	170.73%

12.50%	73.56%	116.59%	147.12%	170.81%
12.60%	73.60%	116.65%	147.19%	170.88%
12.70%	73.63%	116.70%	147.26%	170.96%
12.80%	73.66%	116.75%	147.32%	171.04%
12.90%	73.70%	116.80%	147.39%	171.11%
13.00%	73.73%	116.86%	147.46%	171.19%
13.10%	73.76%	116.91%	147.52%	171.27%
13.20%	73.79%	116.96%	147.59%	171.35%
13.30%	73.83%	117.01%	147.66%	171.42%
13.40%	73.86%	117.07%	147.72%	171.50%
13.50%	73.89%	117.12%	147.79%	171.58%
13.60%	73.93%	117.17%	147.86%	171.66%
13.70%	73.96%	117.23%	147.92%	171.73%
13.80%	73.99%	117.28%	147.99%	171.81%
13.90%	74.03%	117.33%	148.06%	171.89%
14.00%	74.06%	117.38%	148.12%	171.96%
14.10%	74.09%	117.44%	148.19%	172.04%
14.20%	74.13%	117.49%	148.25%	172.12%
14.30%	74.16%	117.54%	148.32%	172.20%
14.40%	74.19%	117.59%	148.39%	172.27%
14.50%	74.23%	117.65%	148.45%	172.35%
14.60%	74.26%	117.70%	148.52%	172.43%
14.70%	74.29%	117.75%	148.59%	172.50%
14.80%	74.33%	117.80%	148.65%	172.58%
14.90%	74.36%	117.86%	148.72%	172.66%
15.00%	74.39%	117.91%	148.78%	172.73%
15.10%	74.43%	117.96%	148.85%	172.81%
15.20%	74.46%	118.01%	148.92%	172.89%
15.30%	74.49%	118.07%	148.98%	172.96%
15.40%	74.52%	118.12%	149.05%	173.04%
15.50%	74.56%	118.17%	149.12%	173.12%
15.60%	74.59%	118.22%	149.18%	173.19%
15.70%	74.62%	118.28%	149.25%	173.27%
15.80%	74.66%	118.33%	149.31%	173.35%
15.90%	74.69%	118.38%	149.38%	173.42%
16.00%	74.72%	118.43%	149.45%	173.50%
16.10%	74.76%	118.49%	149.51%	173.58%
16.20%	74.79%	118.54%	149.58%	173.65%
16.30%	74.82%	118.59%	149.64%	173.73%
16.40%	74.85%	118.64%	149.71%	173.81%
16.50%	74.89%	118.69%	149.78%	173.88%
16.60%	74.92%	118.75%	149.84%	173.96%
16.70%	74.95%	118.80%	149.91%	174.04%
16.80%	74.99%	118.85%	149.97%	174.11%
16.90%	75.02%	118.90%	150.04%	174.19%
17.00%	75.05%	118.96%	150.10%	174.27%

17.10%	75.09%	119.01%	150.17%	174.34%
17.20%	75.12%	119.06%	150.24%	174.42%
17.30%	75.15%	119.11%	150.30%	174.50%
17.40%	75.18%	119.16%	150.37%	174.57%
17.50%	75.22%	119.22%	150.43%	174.65%
17.60%	75.25%	119.27%	150.50%	174.72%
17.70%	75.28%	119.32%	150.57%	174.80%
17.80%	75.32%	119.37%	150.63%	174.88%
17.90%	75.35%	119.42%	150.70%	174.95%
18.00%	75.38%	119.48%	150.76%	175.03%
18.10%	75.41%	119.53%	150.83%	175.11%
18.20%	75.45%	119.58%	150.89%	175.18%
18.30%	75.48%	119.63%	150.96%	175.26%
18.40%	75.51%	119.68%	151.02%	175.33%
18.50%	75.55%	119.74%	151.09%	175.41%
18.60%	75.58%	119.79%	151.16%	175.49%
18.70%	75.61%	119.84%	151.22%	175.56%
18.80%	75.64%	119.89%	151.29%	175.64%
18.90%	75.68%	119.94%	151.35%	175.71%
19.00%	75.71%	120.00%	151.42%	175.79%
19.10%	75.74%	120.05%	151.48%	175.87%
19.20%	75.77%	120.10%	151.55%	175.94%
19.30%	75.81%	120.15%	151.61%	176.02%
19.40%	75.84%	120.20%	151.68%	176.09%
19.50%	75.87%	120.25%	151.74%	176.17%
19.60%	75.91%	120.31%	151.81%	176.25%
19.70%	75.94%	120.36%	151.88%	176.32%
19.80%	75.97%	120.41%	151.94%	176.40%
19.90%	76.00%	120.46%	152.01%	176.47%
20.00%	76.04%	120.51%	152.07%	176.55%
20.10%	76.07%	120.57%	152.14%	176.63%
20.20%	76.10%	120.62%	152.20%	176.70%
20.30%	76.13%	120.67%	152.27%	176.78%
20.40%	76.17%	120.72%	152.33%	176.85%
20.50%	76.20%	120.77%	152.40%	176.93%
20.60%	76.23%	120.82%	152.46%	177.00%
20.70%	76.26%	120.88%	152.53%	177.08%
20.80%	76.30%	120.93%	152.59%	177.16%
20.90%	76.33%	120.98%	152.66%	177.23%
21.00%	76.36%	121.03%	152.72%	177.31%
21.10%	76.39%	121.08%	152.79%	177.38%
21.20%	76.43%	121.13%	152.85%	177.46%
21.30%	76.46%	121.19%	152.92%	177.53%
21.40%	76.49%	121.24%	152.98%	177.61%
21.50%	76.52%	121.29%	153.05%	177.68%
21.60%	76.56%	121.34%	153.11%	177.76%

21.70%	76.59%	121.39%	153.18%	177.83%
21.80%	76.62%	121.44%	153.24%	177.91%
21.90%	76.65%	121.49%	153.31%	177.99%
22.00%	76.69%	121.55%	153.37%	178.06%
22.10%	76.72%	121.60%	153.44%	178.14%
22.20%	76.75%	121.65%	153.50%	178.21%
22.30%	76.78%	121.70%	153.57%	178.29%
22.40%	76.82%	121.75%	153.63%	178.36%
22.50%	76.85%	121.80%	153.70%	178.44%
22.60%	76.88%	121.85%	153.76%	178.51%
22.70%	76.91%	121.91%	153.83%	178.59%
22.80%	76.95%	121.96%	153.89%	178.66%
22.90%	76.98%	122.01%	153.96%	178.74%
23.00%	77.01%	122.06%	154.02%	178.81%
23.10%	77.04%	122.11%	154.09%	178.89%
23.20%	77.08%	122.16%	154.15%	178.96%
23.30%	77.11%	122.21%	154.22%	179.04%
23.40%	77.14%	122.26%	154.28%	179.11%
23.50%	77.17%	122.32%	154.35%	179.19%
23.60%	77.21%	122.37%	154.41%	179.27%
23.70%	77.24%	122.42%	154.48%	179.34%
23.80%	77.27%	122.47%	154.54%	179.42%
23.90%	77.30%	122.52%	154.60%	179.49%
24.00%	77.33%	122.57%	154.67%	179.57%
24.10%	77.37%	122.62%	154.73%	179.64%
24.20%	77.40%	122.67%	154.80%	179.72%
24.30%	77.43%	122.73%	154.86%	179.79%
24.40%	77.46%	122.78%	154.93%	179.87%
24.50%	77.50%	122.83%	154.99%	179.94%
24.60%	77.53%	122.88%	155.06%	180.01%
24.70%	77.56%	122.93%	155.12%	180.09%
24.80%	77.59%	122.98%	155.19%	180.16%
24.90%	77.62%	123.03%	155.25%	180.24%
25.00%	77.66%	123.08%	155.31%	180.31%

Complex Numbers and trigonometric functions cosine and sine

Consider the trigonometric function, cosine.
The cosine function, abbreviated cos, is initially defined as the side adjacent to an acute angle in a right triangle.
For everything that can be said about the cos function, there is a corresponding statement about the sine function, which is abbreviated, sin.
The Pythagorean theorem is expressed by the trigonometric identity,
$(cos(x))^2 + (sin(x))^2 = 1$.
Some special angles are 0 degrees, 30 degrees, 45 degrees, 60 degrees, and 90 degrees.
We identify pi with 180 degrees, pi/2 with 90 degrees, pi/4 with 45 degrees.
Some surprising trigonometric identities show up involving the cos function.
For all angles x, cos (pi/3 – x) + cos(pi/3+x) = cos(x)
The proof of this identity is simple.
One proof requires that we know the identities
for all angles x and y,
Cos(x+y) = cos(x) cos(y) – sin(x) sin(y)
Sin(x+y) = cos(x) sin(y) + sin(x) cos(y)
The proof of the sum of angle identities is also simple if we take as fundamental a basic property of complex numbers.
We define a complex number as a point in the plane, (x,y).
It is convenient to write the complex number (x,y) as x+yi
where i has the property that $i^2 = -1$.

A basic property of complex numbers requires that we define for each complex number a magnitude and angle.
We split the plane into four quadrants. The first quadrant is above the x axis and to the right of the y axis. If x is an angle in the first quadrant, then x is between zero and pi/2.
The second quadrant is counter clockwise of the first quadrant. It is the area left of the y axis and above the x axis.
To define the angle of a point (x,y), draw a line from the point to the origin point. The origin point is the point at x=0 and y=0.
The origin point is (0,0).
That line we draw from (0,0) to the point (x,y) makes a particular angle with the positive x-axis.
That angle is the angle of the complex number (x,y)
The magnitude of the complex number (x,y) is the length of that line from the origin to (x,y). In other words, the magnitude of (x,y) is its distance from the origin.

Now we can state the basic property of complex numbers that allows us to discover many interesting properties of the trigonometric functions.

That property is: To multiply two complex numbers together, add their angles, and multiply their magnitudes.

Suppose we have a complex number (x,y). What, in terms of x and y, is the magnitude of (x,y). What in terms of x and y is the angle of (x,y)?

Draw the right triangle consisting of the line segment from the origin to (x,y), the line segment straight down from (x,y) to the x-axis, and the line segment along the x-axis from that intersection point, back to the origin.

The length of the line segment along the x-axis is x. The length of the line segment parallel to the y axis from (x,y) down to the x-axis, is y.

Let r be the distance of (x,y) from the origin. The hypotenuse of this right triangle is r. Apply the Pythagorean theorem to get that $r^2 = x^2 + y^2$.

Thus the magnitude of (x,y) = sqrt(x^2+y^2), where sqrt means square root.

What is the angle of (x,y).

The angle of (x,y) is the angle that the hypotenuse of our triangle makes with the x-axis. Suppose we name that angle, t.

In terms of the right triangle we drew, we define the cos(t) to be equal to x/r, and we define the sin(t) to be y/r.

Consider two complex numbers (x_1,y_1) and (x_2,y_2).

Let r_1 be the magnitude of (x_1,y_1) and r_2 be the magnitude of (x_2,y_2).

Let t_1 be the angle of (x_1,y_1), and t_2 be the angle of (x_2,y_2).

$\cos(t_1) = x_1/r_1$.

$\sin(t_1) = y_1/r_1$.

$\cos(t_2) = x_2/r_2$.

$\sin(t_2) = y_2/r_2$.

Rewrite these equations as

$x_1 = r_1 \cos(t_1)$

$y_1 = r_1 \sin(t_1)$

$x_2 = r_2 \cos(t_2)$

$y_2 = r_2 \sin(t_2)$.

Now let's use the standard complex number notation for the numbers (x_1,y_1) and (x_2,y_2).

$(x_1,y_1) = x_1 + i\, y_1 = r_1(\cos(t_1) + i \sin(t_1))$

$(x_2,y_2) = x_2 + i\, y_2 = r_2(\cos(t2) + i \sin(t_2))$

Now lets multiply the complex numbers (x_1,y_1) and (x_2,y_2).

We apply the basic property of complex numbers that to multiply complex numbers you add their angles and multiply their magnitudes.

$(x_1,y_1) = r_1 (\cos(t_1) + i \sin(t_1))$

$(x_2,y_2) = r_2 (\cos(t_2) + i \sin(t_2))$

The product $(x1,y_1) * (x_2,y_2)$ is

$r_1 r_2 (\cos(t_1+t_2) + i \sin(t_1+t_2))$

We can also calculate the product another way and compare the results.

$(x_1,y_1) = x_1 + y_1 i$

$(x_2,y_2) = x_2 + y_2 i$

$(x_1,y_1)*(x_2,y_2) = (x_1+y_1 i)(x_2 + y_2 i)$

$= (x_1 (x_2 + y_2 i) + (y_1 i) (x_2 + y_2 i)$

$= (x_1 x_2 + x_1 y_2 i) + (y_1 x_2 i + y_1 y_2 i^2)$

$= (x_1 x_2 - y_1 y_2) + (x_1 y_2 + y_1 x_2) i$

Compare this final product with our previous derivation.

$r_1 r_2 (\cos(t_1+t_2) + i \sin(t_1+t_2))$

We conclude that

$r_1\,r_2\,(\cos(t_1+t_2) + i\,\sin(t_1+t_2) = (x_1\,x_2 - y_1\,y_2) + (x_1\,y_2 + y_1\,x_2)\,i$
Divide through by the magnitudes, and we get
$(\cos(t_1+t_2) + i\,\sin(t_1+t_2)$
$= ((x_1/r_1)(x_2/r_2)-(y_1/r_1)(y_2/r_2))+((x_1/r_1)(y_2/r_2)+(y_1/r_1)(x_2/r_2))$

which yields
$\cos(t_1+t_2)=\cos(t_1)\cos(t_2)-\sin(t_1)\sin(t_2)$ and
$\sin(t_1+t_2) = \cos(t_1)\,\sin(t_2) + \sin(t_1)\,\cos(t_2)$.

Finally we apply these two formula to
$\cos(pi/3 - t) + \cos(pi/3 + t)$.

$\cos(pi/3 -t) = \cos(pi/3)\,\cos(t) + \sin(pi/3)\,\sin(t)$
$\cos(pi/3+t) = \cos(pi/3)\,\cos(t) - \sin(pi/3)\,\sin(t)$.
Note that when t was negative that the sign of sin also flipped.
$\sin(-t) = -\sin(t)$.
Add the two cos equations.
$\cos(pi/3 - t) + \cos(pi/3+t) = 2\,\cos(pi/3)\,\cos(t)$.

What is the value of $\cos(pi/3)$. It is ½.

$\cos(pi/3 -t) + \cos(pi/3 + t) = 2(1/2)\,\cos(t) = \cos(t)$.

Triangle Puzzle

> Let a,b,c be real number that satisfy :
> cos(a)+cos(b)+cos(c)=0
> and
> sin(a)+sin(b)+sin(c)=0
>
> Show that :
> cos(2a)+cos(2b)+cos(2c)=0
> and
> sin(2a)+sin(2b)+sin(2c)=0
>

This one is easy. It is solved by applying Euler's identity.
$e^{it} = \cos(t) + i \sin(t)$. An immediate corollary of Euler's identity is that
$\cos(2t) + i \sin(2t) = e^{i(2t)} = (e^{it})^2 = (\cos(t) + i \sin(t))^2$
Because $(\cos(t))^2 + (\sin(t))^2 = 1$, all the complex numbers of form e^{it} lie on the unit circle.

$\cos(a) + \cos(b) + \cos(c) = 0$
$\sin(a) + \sin(b) + \sin(c) = 0$

$i[\sin(a) + \sin(b) + \sin(c)] = 0$

$[\cos(a) + i \sin(a)] + [\cos(b) + i \sin(b)] + [\cos(c) + i \sin(c)] = 0$

Let $z_1 = \cos(a) + i \sin(a)$

$z_2 = \cos(b) + i \sin(b)$

$z_3 = \cos(c) + i \sin(c)$

z_1, z_2, z_3 are complex numbers which lie on the unit circle.
$z_1 + z_2 + z_3 = 0$

$z_1 + z_2 = -z_3$

$z_1/z_3 + z_2/z_3 = -1$

Let $w_1 = z_1/z_3$

$w_2 = z_2/z_3$

w_1 and w_2 are on the unit circle because z_1, z_2, and z_3 are on the unit circle.
$w_1 + w_2 = -1$

Because $w_1 + w_2$ is a real number,

w_1 and w_2 are complex conjugates on the unit circle. The product of conjugate complex numbers is a real number which is the sum of two squares.

$w_1 * w_2$ is the sum of squares and is a positive real number on the intersection of the x-axis with the unit circle.

$w_1 * w_2 = 1$
$(w_1 + w_2)^2 = (-1)^2 = 1$

$w_1^2 + 2 * w_1 * w_2 + w_2^2 = 1$

$w_1^2 + 2 + w_2^2 = 1$

$w_1^2 + w_2^2 = -1$

$w_1^2 + w_2^2 + 1 = 0$

$z_3^2 (w_1^2 + w_2^2 + 1) = 0$

$(z_3 * w_1)^2 + (z_3 * w_2)^2 + z_3^2 = 0$

$w_1 = z_1/z_3; \quad w_2 = z_2/z_3;$

$z_3 * w_1 = z_1; \quad z_3 * w_2 = z_2$

$z_1^2 + z_2^2 + z_3^2 = 0$

$z_1 = \cos(a) + i \sin(a); \quad z_2 = \cos(b) + i \sin(b); \quad z_3 = \cos(c) + i \sin(c)$

$z_1^2 = \cos(2a) + i \sin(2a); \quad z_2^2 = \cos(2b) + i \sin(2b); \quad z_3^2 = \cos(2c) + i \sin(2c)$

$z_1^2 + z_2^2 + z_3^2 = 0$

$[\cos(2a) + i \sin(2a)] + [\cos(2b) + i \sin(2b)] + [\cos(2c) + i \sin(2c)] = 0$

$[\cos(2a) + \cos(2b) + \cos(2c)] + i [\sin(2a) + \sin(2b) + \sin(2c)] = 0$

$\cos(2a) + \cos(2b) + \cos(2c) = 0$
$\sin(2a) + \sin(2b) + \sin(2c) = 0$

QED

Difference of two squares and sum of two squares

It is generally well known that if $z_1 = a_1^2 + b_1^2$ and $z_2 = a_2^2 + b_2^2$,
we can calculate the product $z_1 * z_2$ as the sum of two squares.

$z_1 * z_2 = (a_1 a_2 - b_1 b_2)^2 + (a_1 b_2 + b_1 a_2)^2 = (a_1 a_2 + b_1 b_2)^2 - (a_1 b_2 - b_1 a_2)^2$

$5 = 2^2 + 1^2$
$13 = 3^2 + 2^2$

$5 * 13 = (2*3 + 1 * 2)^2 + (2 * 2 - 1 * 3)^2 = 8^2 + 1^2$
$5 * 13 = (2*3 - 1 * 2)^2 + (2 * 2 + 1 * 3)^2 = 4^2 + 7^2$

It is also true that if $w_1 = a_1^2 - b_1^2$ and $w_2 = a_2^2 - b_2^2$, then we can calculate the product $w_1 * w_2$ as the difference of two squares.

$w_1 * w_2 = (a_1 a_2 + b_1 b_2)^2 - (a_1 b_2 + b_1 a_2)^2 = (a_1 a_2 - b_1 b_2)^2 - (a_1 b_2 - b_1 a_2)^2$

$3 = 2^2 - 1^2$
$5 = 3^2 - 2^2$
$3 * 5 = (2*3 + 1 * 2)^2 - (2*2+1*3)^2 = 8^2 - 7^2$
$3 * 5 = (2*3 - 1*2)^2 - (2*2 - 1*3)^2 = 4^2 - 1^2$

One immediate application of the product of two squares theorem is that we can easily find the general formula for squares that are the sum of two squares.

Let $z1 = z2 = a^2 + b^2$ be the sum of two squares.

Apply the formula

$z_1 * z_2 = (a_1 a_2 - b_1 b_2)^2 + (a_1 b_2 + b_1 a_2)^2. = (a_1 a_2 + b_1 b_2)^2 + (a_1 b_2 - b_1 a_2)^2$

Result is

$z^2 = (a^2 - b^2)^2 + (2 a b)^2 = (a^2 + b^2)^2 + 0^2$

Thus we see that whenever z is the sum of two squares, so is z^2 a sum of two squares.

The converse is also true. If z^2 is the sum of two squares, so is z.

Suppose we have a number that is both the sum of two squares and the difference of two squares.

$4(a_1^2 + b_1^2)(a_2^2 + b_2^2) = 4(a_1 a_2 - b_1 b_2)^2 + 4(a_1 b_2 + b_1 a_2)^2 = (a_1^2 + b_1^2 + a_2^2 + b_2^2)^2 - (a_1^2 + b_1^2 - a_2^2 - b_2^2)^2$
Then we write this as:
$4(a_1^2 + b_1^2)(a_2^2 + b_2^2)$
$= 4(a_1 a_2 - b_1 b_2)^2 + 4(a_1 b_2 + b_1 a_2)^2 + (a_1^2 + b_1^2 - a_2^2 - b_2^2)^2 = (a_1^2 + b_1^2 + a_2^2 + b_2^2)^2$

Enhancing Fermat Factoring

How would we most quickly use Fermat Factoring to determine two positive integers which, when multiplied together, make 1000009?

If $t^2 - s^2 = 1000009$, then $(t-s) * (t+s) = 1000009$.

We can search for values of t and s that solve this equation.

Once we have found the correct value of t, we can then quickly determine the values of $s = sqrt(t^2 - 1000009)$.

How can we most quickly determine the value of t?

We know that $t > sqrt(1000009)$ because

$t = sqrt(1000009 + s^2)$.

What other constraints can we place on t and s to help us quickly find their values?

$t > sqrt(1000009)$.
$t > 1000$.

$2 * 3 * 5 * 7 * 11 * 13 = 30030$.

$t^2 - s^2 = 1000009$ mod 30030.

Since each prime factor cuts in half the number of t-values and s-values that we need to look at, we have improved the naïve Fermat factoring algorithm by a factor of 64. However, there are other factoring algorithms, such as the Pollard Rho algorithm, which are much better than even this enhanced Fermat factoring algorithm.

We can implement an algorithm as follows:

Generate the possible t values and s values mod 30030.

Start with the smallest permitted t-value that is $>$ sqrt(1000009) and the smallest permitted s value.

Calculate $t^2 - s^2$.
If $(t^2 - s^2)$ is greater than 1000009, increase s by taking the next largest permitted s value.
If $(t^2 - s^2)$ is less than 1000009, increase t by taking the next largest permitted t value.
If $(t^2 - s^2)$ is equal to 1000009, return factors (t-s) and (t+s).
Loop until factors are found.

Description of my version of the Pollard Rho Factorization algorithm

We wish to factor some positive integer, z.

First test if z is prime. The Pollard Rho factorization algorithm will fail if z is a prime number.

First I describe the algorithm, then I discuss the reason that it works.

Pick some polynomial that is relatively easy to evaluate.
A popular choice is the polynomial $P(x) = x^2 + 1$.

Initialize 6 variables as follows.
     RhoA = 5
     RhoB = 5
     Rhop1 = 1 # to hold accumulated product
     Rhodiff = 1 # to hold the difference, (RhoA – RhoB) mod z
     RhoLim = 1 # to hold how many times do only the multiplication mod z.
               # Each time the RhoLim limit is reached by RhoCount, RhoLim is doubled.
     RhoCount = 0   # to hold how the count up to RhoLim.
               # Each time the RhoLim limit is reached by RhoCount, reset RhoCount to 0.

Loop indefinitely.

k = 0
     while True:
          k = k + 1

# Loop RHO algorithm for polynomial x^2 + 1
          Rhop = Rhop1 + zero   #   Copy value in temporary variable Rhop1 to variable Rhop.
          RhoCount = RhoCount + 1   #Increment RhoCount.
          RhoA = (RhoA * RhoA + 1)%z   #Calculate polynomial (RhoA^2 + 1).
                                        # store back in RhoA.
          RhoB = (RhoB * RhoB + 1)%z   #calculate RhoB = (RhoB^2+1)^2 + 1.
          RhoB = (RhoB * RhoB + 1)%z
          RhoDiff = (RhoA - RhoB)%z   # Calculate difference mod z.
                                        # RhoDiff = (RhoA – RhoB) mod z
          Rhop1 = (Rhop * RhoDiff)%z   # Multiply accumulated product my RhoDiff mod z.
                                        # Store in temporary variable Rhop1.
          if Rhop1 == 0:   # Means that Rhop1 was zero mod q for all q which divide z.
             Rhox = GCD(z,Rhop)   # Take GCD of z and the previous accumulated product.
             if 1 < Rhox < z: # If GCD is > 1 and < z, then factors have been found.
                return [Rhox,z//Rhox,k,"Pollard Rho x^2+1 All prime Factors"]
   # return factor1, factor2, number of iterations, choice of polynomial,
   # and reason algorithm terminated.

```
        if RhoCount > RhoLim:   #   Test if we are ready to take GCD.
            RhoLim = RhoLim * 2   #      If current limit has been reached,  double it.
            RhoCount = 0          #  Reset counter for limit back to zero.
            Rhox = GCD(z,Rhop1)      #  Take GCD with previous accumulated product.
            if 1 < Rhox < z:             #  If factors found, return them.
                return [Rhox,z//Rhox,k,"Pollard Rho, x^2 + 1, First factor check"]
```

An illustration of the Pollard Rho algorithm will make clear why it works.

Suppose we wish to factor 341 by Pollard Rho.

We keep track of our variable values as follows:

RhoA = 5
RhoB = 5
Rhop1 = 1
Rhodiff = 1
RhoLim = 1
RhoCount = 0

k = 0
k = 1
Rhop = 5 + 0 = 5
RhoCount = 1
RhoA = 26
RhoB = 26
RhoB = 336 # Note that 336 = (341 − 5)
RhoDiff = 31
Rhop1 = 31
Rhop1 is not zero mod 341, so skip to check if ready to take GCD.
Compare RhoCount to RhoLim.
RhoCount is not greater than RhoLim.
Continue main loop.
Rhop = 31
RhoCount = 2
RhoA = 336
RhoB = 336
RhoDiff = 0
Rhop1 = 0
Take GCD of 341 with previous accumulated product.
GCD(341,31) = 31.
We have found a factor.

The factors of 341 are 11 and 31.

In mod 11, the generated values of RhoA are 5, 4, 6,4, ...
In mod 11, the generated values of RhoB are 5, 6, 6, 6, ...

In mod 31, the generated values of RhoA are 5, 26, 26, 26, ...
In mod 31, the generated values of RhoB are 5, 26, 26, 26, ...

At the second iteration, the difference in both mod 11 and mod 31 is zero.

The difference in both mod 11 and mod 31 is zero if and only if the difference mod 341 is zero.

If the difference in one of the mods had been zero, but not zero in the other mod,
then we would have found a factor the next time we checked the GCD.

The first GCD check is at the second iteration.
That is we go through the multiply only loop 1 time for the first time.

Then we double the length of the multiply only loop.
The length of the multiply only loop goes from 1 to 2 to 4 to 8 to....
until a factor is found.

Factoring numbers that are strong pseudo primes to some small base

The Rabin miller probable prime test is described on the wiki page
http://en.wikipedia.org/wiki/Miller%E2%80%93Rabin_primality_test
Suppose z is an odd positive integer to be factored by the probably prime test method.
Calculate d and s such that
$z = 1 + d * 2^s$.
Let k be, in sequence, 2,3,4,…
For each k, calculate
$H_2 = k^d$ mod z
Repetitively calculate
$H_1 = H^2$
$H_2 = H_1 * H_1$ mod z
until either you have done this s times or $H_2 = 1$ mod z,
whichever comes firsts.

If $H_2 = 1$, then z might be prime,
or it might have factors in common with $(H_1 - 1)$.

Example: Factor 341 by this method.
Calculate d and s such that
$341 = 1 + d * 2^s$.

d = 85, s = 2 because
$341 = 1 + 85 * 2^2$.

Let's first try base 2.

Calculate
$H_2 = 2^{85}$ mod 341 = 32 mod 341.

$H_1 = 32$ mod 341; $H_2 = 32*32$ mod 341 = 1

We have reached 1 before trying it 2 times. So 341 might some factor in common with 32-1 = 31.

31 does divide 341. We have factored 341.

Sum of three squares = sum of three squares

If
$t_1 = a_1*d_1 + a_2*d_2 - b_1*c_1 - b_2*c_2;$

$s_1 = a_1*d_1 - a_2*d_2 - b_1*c_1 + b_2*c_2;$

$t_2 = a_1*b_2 + c_1*a_2 + b_1*d_2 + d_1*c_2;$

$s_2 = a_1*b_2 - c_1*a_2 + b_1*d_2 - d_1*c_2;$

$t_3 = a_1*a_2 + c_1*b_2 + b_1*c_2 + d_1*d_2;$

$s_3 = a_1*a_2 - c_1*b_2 + b_1*c_2 - d_1*d_2;$

then

$t_1*t1 + t_2*t_2 + s_3*s_3 = s_1*s_1 + s_2*s_2 + t_3*t_3$

Illustration:

Set

$t_1 = 3*19 + 5*23 - 7*13 - 11*17 = -106$

$s_1 = 3*19 - 5*23 - 7*13 + 11*17 = 38$

$t_2 = 3*11 + 13*5 + 7*23 + 19*17 = 582$

$s_2 = 3*11 - 13*5 + 7*23 - 19*17 = -194$

$t_3 = 3*a2 + 13*11 + 7*17 + 19*23 = 714$

$s_3 = 3*5 - 13*11 + 7*17 - 19*23 = -446$

Then

$t_1*t_1 + t_2*t_2 + s_3*s_3 = s_1*s_1 + s_2*s_2 + t_3*t_3$
$= 11236 + 338724 + 198916 = 1444 + 37636 + 509796 = 548876$

Linear transformation of Pythagorean triples into Pythagorean triples

First I learned this special linear transformation of Pythagorean Triple into Pythagorean Triple.

[Theorem 1] If $x^2 + y^2 = z^2$,
then $(x - 2y + 2z)^2 + (2x - y + 2z)^2 = (2x - 2y + 3z)^2$

Proof:

$(x - 2y + 2z)^2 + (2x - y + 2z)^2 - (2x - 2y + 3z)^2$

$= (x^2 + 4y^2 + 4z^2 - 4xy + 4xz - 8yz)$
$\;+(4x^2 + y^2 + 4z^2 - 4xy + 8xz - 4yz)$
$+(-4x^2 - 4y^2 - 9z^2 + 8xy - 12xz + 12yz)$

$= x^2 + y^2 - z^2$

$= 0$

Then I looked for, and found this generalization.

[Theorem 2] For any integer q,
If $x^2 + y^2 = z^2$
Then
$((2q^2 - 1)x - (2q)y + (2q^2)z)^2$
$+((2q)x - (1)y + (2q)z)^2$
$=((2q^2)x - (2q)y + (2q^2 + 1)z)^2$

Proof:
$((2q^2 - 1)x - (2q)y + (2q^2)z)^2$
$+((2q)x - (1)y + (2q)z)^2$
$-((2q^2)x - (2q)y + (2q^2 + 1)z)^2$
$= ((2q^2 - 1)^2 + (2q)^2 - (2q^2)^2) x^2$
$+((2q)^2 + 1 - (2q)^2) y^2$
$+((2q^2)^2 + (2q)^2 - (2q^2 + 1)^2) z^2$
$+(-2(2q)(2q^2 - 1) - 2(2q) + 2(2q)(2q^2)) xy$
$+(2(2q^2 - 1)(2q^2) + 2(2q)(2q) - 2(2q^2)(2q^2 + 1)) xz$
$+(-2(2q)(2q^2) - 2(2q) + 2(2q)(2q^2 + 1)) yz$
$= (4q^4 - 4q^2 + 1 + 4q^2 - 4Q^4)x^2$
$+ (4q^2 + 1 - 4q^2) y^2$
$+ (4q^4 + 4q^2 - 4q^4 - 4q^2 - 1)z^2$
$+ (-8q^3 + 4q - 4q + 8q^3)xy$
$+ (8q^4 - 4q^2 + 8q^2 - 8q^4 - 4q^2) xz$

+ $(-8q^3 -4q + 8q^3 + 4q)$ yz
= $x^2 + y^2 - z^2 + 0$ xy + 0 xz + 0 yz
= $x^2 + y^2 - z^2$
= 0

To confirm that [Theorem 2] is a generalization of [Theorem 1],
set q = 1 in [Theorem 2] to get [Theorem 1].

Finally, I searched for, and found, the most complete possible generalization.

[Theorem 3] If $x^2 + y^2 = z^2$,
and
q1 = $(1/2)(a^2 - b^2 - c^2 + d^2)$,
q2 = (a c - b d),
q3 = $(1/2)(a^2 - b^2 + c^2 - d^2)$,
q4 = (a b - c d),
q5 = (a d + b c),
q6 = (a b + c d),
q7 = $(1/2)(a^2 + b^2 - c^2 - d^2)$,
q8 = (a c + b d),
q9 = $(1/2)(a^2 + b^2 + c^2 + d^2)$,

then

(q1 x + q2 y + q3 z)^2 + (q4 x + q5 y + q6 z)^2 = (q7 x + q8 y + q9 z)^2.

Proof:

If there is a common factor, J, in x, y, and z,
then that same factor, J, is common to

(q1 x + q2 y + q3 z),   (q4 x + q5 y + q6 z), and (q7 x + q8 y + q9 z).

Thus, without loss of generality, we may assume that x, y, and z, have no factor in common.

Then we may write

x = $(t^2 - s^2)$,
y = (2 t s)
z = $(t^2 + s^2)$

Calculate

(q1 x + q2 y + q3 z)^2 + (q4 x + q5 y + q6 z)^2 - (q7 x + q8 y + q9 z)^2.

$= (q1\,(t^2 - s^2) + q2\,(2\,t\,s) + q3\,(t^2 + s^2))^2 + (q4\,(t^2 - s^2) + q5\,(2\,t\,s) + q6\,(t^2 + s^2))^2$
$- (q7\,(t^2 - s^2) + q8\,(2\,t\,s) + q9\,(t^2 + s^2))^2.$

$= ((1/2)(a^2 - b^2 - c^2 + d^2)\,(t^2 - s^2) + (a\,c - b\,d)\,(2\,t\,s)$
$+ (1/2)(a^2 - b^2 + c^2 - d^2)\,(t^2 + s^2))^2$
$+ ((a\,b - c\,d)\,(t^2 - s^2) + (a\,d + b\,c)\,(2\,t\,s) + (a\,b + c\,d)\,(t^2 + s^2))^2$
$- ((1/2)(a^2 + b^2 - c^2 - d^2)\,(t^2 - s^2) + (a\,c + b\,d)\,(2\,t\,s)$
$+ (1/2)(a^2 + b^2 + c^2 + d^2)\,(t^2 + s^2))^2$

$= ((a^2 - b^2)\,t^2 + 2\,(a\,c - b\,d)\,t\,s + (c^2 - d^2)\,s^2)^2$
$+ (2\,a\,b\,t^2 + 2\,(a\,d + b\,c)\,t\,s + 2\,c\,d\,s^2)^2$
$- ((a^2 + b^2)\,t^2 + 2\,(a\,c + b\,d)\,t\,s + (c^2 + d^2)\,s^2)^2$

$= (\,(a^2 - b^2)^2 + (2\,a\,b)^2 - (a^2 + b^2)^2\,)\,t^4$
$+ (4\,(a^2 - b^2)\,(a\,c - b\,d) + 8\,a\,b\,(a\,d + b\,c) - 4(a^2 + b^2)\,(a\,c + b\,d)\,)\,t^3\,s$
$+ ((4\,(a\,c - b\,d)^2 + 2\,(a^2 - b^2)\,(c^2 - d^2) + 4\,(a\,d + b\,c)^2 + 8\,a\,b\,c\,d - 4\,(a\,c + b\,d)^2$
$- 2\,(a^2 + b^2)\,(c^2 + d^2)\,)\,t^2\,s^2$
$+ (\,4\,(a\,c - b\,d)\,(c^2 - d^2) + 8\,(a\,d + b\,c)\,c\,d - 4\,(a\,c + b\,d)\,(c^2 + d^2)\,)\,t\,s^3$
$+ ((c^2 - d^2)^2 + (2\,c\,d)^2 - (c^2 + d^2)^2)\,s^4$

$= 0\,t^4$
$+ (4\,a^3\,c - 4\,a^2\,b\,d - 4\,a\,b^2\,c + 4\,b^3\,d + 8\,a^2\,b\,d + 8\,a\,b^2\,c$
$- 4\,a^3\,c - 4\,a^2\,b\,d - 4\,a\,b^2\,c - 4\,b^3\,d)\,t^3\,s$
$+ (4\,a^2\,c^2 - 8\,a\,b\,c\,d + 4\,b^2\,d^2 + 2\,a^\wedge\,c^2 - 2\,a^2\,d^2 - 2\,b^2\,c^2 + 2\,b^\wedge d^2$
$+ 4\,a^2\,d^2 + 8\,a\,b\,c\,d + 4\,b^2\,c^2$
$+ 8\,a\,b\,c\,d - 4\,a^2\,c^2 - 8\,a\,b\,c\,d - 4\,b^2\,d^2 - 2\,a^2\,c^2 - 2\,a^2\,d^2 - 2\,b^2\,c^2$
$- 2\,b^2\,d^2\,)\,t^2\,s^2$
$+ (4\,a\,c^3 - 4\,a\,c\,d^2 - 4\,b\,c^2\,d + 4\,b\,d^3 + 8\,a\,c\,d^2 + 8\,b\,c^2\,d - 4\,a\,c^3 - 4\,a\,c\,d^2$
$- 4\,b\,c^2\,d - 4\,b\,d^3)\,t\,s^3$
$+ 0\,s^4$

$= 0\,t^4$
$+ 0\,t^3\,s$
$+ 0\,t^2\,s^2$
$+ 0\,t\,s^3$
$+ 0\,s^4$

$= 0$

To confirm that [Theorem 3] is a generalization of [Theorem 2],

set a = (-2q), b = (-1), c = 1, and d = 0.

Then the definitions for q1 through q9,

q1 = (1/2)(a^2 - b^2 - c^2 + d^2),
q2 = (a c - b d),
q3 = (1/2)(a^2 - b^2 + c^2 - d^2),
q4 = (a b - c d),
q5 = (a d + b c),
q6 = (a b + c d),
q7 = (1/2)(a^2 + b^2 - c^2 - d^2),
q8 = (a c + b d),
q9 = (1/2)(a^2 + b^2 + c^2 + d^2),

become the identities

(1/2)((-2 q)^2 - (-1)^2 - 1) = (2 q^2 - 1)
(-2 q) = - 2 q
(1/2)((-2 q)^2 - (-1)^2 + 1) = 2 q^2
((-2 q) (-1)) = 2 q
(-1) = -1
((-2 q) (-1)) = 2 q
(1/2)((-2 q)^2 + (-1)^2 - 1) = 2 q^2
(-2 q) = -2q
(1/2)((-2 q)^2 + (-1)^2 + 1) = (2q^2 + 1)

Also, note that

If x^2 + y^2 = z^2 and a^2 + b^2 = c^2,
then
[1] (a x – b y)^2 + (a y + b x)^2 = (c z)^2
[2] (a x + b y)^2 + (a y – b x)^2 = (c z)^2
[3] (b y)^2 + (az + cx)^2 = (cz + ax)^2
[4] (b y)^2 + (az - cx)^2 = (cz - ax)^2
[5] (a y)^2 + (b z + c x)^2 = (c z + b x)^2
[6] (a y)^2 + (b z - c x)^2 = (c z - b x)^2
[7] (b x)^2 + (a z + c y)^2 = (c z + a y)^2
[8] (b x)^2 + (a z - c y)^2 = (c z - a y)^2
[9] (a x)^2 + (b z + c y)^2 = (c z + b y)^2
[10] (a x)^2 + (b z - c y)^2 = (c z - b y)^2

120 120th roots of 1

Sequence
Number   120th Root of 1

1: ((((sqrt(30)+sqrt(2))*(sqrt(10-2*sqrt(5))+2)
+ (sqrt(10)+sqrt(6))*(sqrt(10-2*sqrt(5))-2))/32)
+ i * (((2*sqrt(30)+2*sqrt(10)-2*sqrt(6)-2*sqrt(2)) + (sqrt(2)-sqrt(6)+sqrt(10)
-sqrt(30))*sqrt(10-2*sqrt(5)))/32)

2: ((sqrt(15)+sqrt(3)+sqrt(10-2*sqrt(5)))/8)
+ i * ((sqrt(30-6*sqrt(5)) - 1 -sqrt(5))/8)

3: (sqrt(2) * ((1+sqrt(5)) + sqrt(10 - 2 * sqrt(5))) /8)
+ i * ((sqrt(10)+sqrt(2) -sqrt(20-4*sqrt(5)))/8)

4: (((sqrt(15)+sqrt(3))*sqrt(10-2*sqrt(5)) + 2*sqrt(5)-2)/16)
+ i * ((2*sqrt(3)*(1-sqrt(5))+(1+sqrt(5))*sqrt(10-2*sqrt(5)))/16)

5: ((sqrt(6)+sqrt(2))/4) + i * ((sqrt(6)-sqrt(2))/4)

6: ((1+sqrt(5))*sqrt(10-2*Sqrt(5))/8) + i * ((-1+sqrt(5))/4)

7: ((sqrt(30)+sqrt(10)+sqrt(6)+sqrt(2)+(sqrt(6)-sqrt(2))*sqrt(10-2*sqrt(5)))/16)
+ i * ((2*(sqrt(15-3*sqrt(5)) + sqrt(5-sqrt(5))) - sqrt(30) + sqrt(10) - sqrt(6) + sqrt(2))/16)

8: ((sqrt(30-6*sqrt(5)) + sqrt(5) + 1)/8) + i * ((sqrt(15)+sqrt(3)-sqrt(10-2*sqrt(5)))/8)

9: (((1+sqrt(5))*sqrt(10-2*sqrt(5)) + 2 * sqrt(5) - 2) * sqrt(2)/16)
+ i * (((1+sqrt(5))*sqrt(10-2*sqrt(5)) - 2 * sqrt(5) + 2) * sqrt(2)/16)

10: (Sqrt(3)/2) + i * (1/2)

11: (((sqrt(30)+sqrt(10)+sqrt(6)+sqrt(2))*sqrt(10-2*sqrt(5))
-2*(sqrt(30)-sqrt(10)-sqrt(6)+sqrt(2)))/32)
+ i * (((sqrt(30)-sqrt(10)+sqrt(6)-sqrt(2))*sqrt(10-2*sqrt(5))
+2*(sqrt(30)+sqrt(10)-sqrt(6)-sqrt(2)))/32)

12: ((1+sqrt(5))/4) + i * (sqrt(10-2*sqrt(5))/4)

13: ((sqrt(30)+sqrt(10)+sqrt(6)+sqrt(2)+(sqrt(6)-sqrt(2))*sqrt(10-2*sqrt(5)))/16
- (SQRT(10)+SQRT(2) -SQRT(20-4*SQRT(5)))/8)
+ i * ((sqrt(2)) * (((1+sqrt(3))*(1 + sqrt(5)) +(1   - sqrt(3)) * sqrt(10-2*sqrt(5)))   ) /16)

14: $(((((\text{sqrt}(30)+\text{sqrt}(2))*(\text{sqrt}(10-2*\text{sqrt}(5))+2)$
$+ (\text{sqrt}(10)+\text{sqrt}(6))*(\text{sqrt}(10-2*\text{sqrt}(5))-2))/32 + ((2*\text{sqrt}(30)+2*\text{sqrt}(10)-2*\text{sqrt}(6)-2*\text{sqrt}(2))$
$+ (\text{sqrt}(2)-\text{sqrt}(6)+\text{sqrt}(10)-\text{sqrt}(30))*\text{sqrt}(10-2*\text{sqrt}(5)))/32) * (\text{sqrt}(2)/2))$
$+ i * (((((\text{sqrt}(30)+\text{sqrt}(2))*(\text{sqrt}(10-2*\text{sqrt}(5))+2)$
$+ (\text{sqrt}(10)+\text{sqrt}(6))*(\text{sqrt}(10-2*\text{sqrt}(5))-2))/32 - ((2*\text{sqrt}(30)+2*\text{sqrt}(10)-2*\text{sqrt}(6)-2*\text{sqrt}(2))$
$+ (\text{sqrt}(2)-\text{sqrt}(6)+\text{sqrt}(10)-\text{sqrt}(30))*\text{sqrt}(10-2*\text{sqrt}(5)))/32) * (\text{sqrt}(2)/2))$

15: $(\text{sqrt}(2)/2) + i * (\text{sqrt}(2)/2)$

16: $(((((\text{sqrt}(30)+\text{sqrt}(2))*(\text{sqrt}(10-2*\text{sqrt}(5))+2)$
$+ (\text{sqrt}(10)+\text{sqrt}(6))*(\text{sqrt}(10-2*\text{sqrt}(5))-2))/32 - ((2*\text{sqrt}(30)+2*\text{sqrt}(10)-2*\text{sqrt}(6)-2*\text{sqrt}(2))$
$+ (\text{sqrt}(2)-\text{sqrt}(6)+\text{sqrt}(10)-\text{sqrt}(30))*\text{sqrt}(10-2*\text{sqrt}(5)))/32) * (\text{sqrt}(2)/2))$
$+ i * (((((\text{sqrt}(30)+\text{sqrt}(2))*(\text{sqrt}(10-2*\text{sqrt}(5))+2) + (\text{sqrt}(10)+\text{sqrt}(6))*(\text{sqrt}(10-2*\text{sqrt}(5))-2))/32$
$+ ((2*\text{sqrt}(30)+2*\text{sqrt}(10)-2*\text{sqrt}(6)-2*\text{sqrt}(2))$
$+ (\text{sqrt}(2)-\text{sqrt}(6)+\text{sqrt}(10)-\text{sqrt}(30))*\text{sqrt}(10-2*\text{sqrt}(5)))/32) * (\text{sqrt}(2)/2))$

17: $((\text{sqrt}(2)) * (((1+\text{sqrt}(3))*(1 + \text{sqrt}(5)) +(1 - \text{sqrt}(3)) * \text{sqrt}(10-2*\text{sqrt}(5)))) /16)$
$+ i * ((\text{sqrt}(30)+\text{sqrt}(10)+\text{sqrt}(6)+\text{sqrt}(2)+(\text{sqrt}(6)-\text{sqrt}(2))*\text{sqrt}(10-2*\text{sqrt}(5)))/16$
$- (\text{SQRT}(10)+\text{SQRT}(2) -\text{SQRT}(20-4*\text{SQRT}(5)))/8$

18: $(\text{sqrt}(10-2*\text{sqrt}(5))/4) + i * ((1+\text{sqrt}(5))/4)$

19: $(((\text{sqrt}(30)-\text{sqrt}(10)+\text{sqrt}(6)-\text{sqrt}(2))*\text{sqrt}(10-2*\text{sqrt}(5))$
$+2*(\text{sqrt}(30)+\text{sqrt}(10)-\text{sqrt}(6)-\text{sqrt}(2)))/32)$
$+ i * (((\text{sqrt}(30)+\text{sqrt}(10)+\text{sqrt}(6)+\text{sqrt}(2))*\text{sqrt}(10-2*\text{sqrt}(5))$
$-2*(\text{sqrt}(30)-\text{sqrt}(10)-\text{sqrt}(6)+\text{sqrt}(2)))/32)$

20: $(1/2) + i * (\text{sqrt}(3)/2)$

21: $(((1+\text{sqrt}(5))*\text{sqrt}(10-2*\text{sqrt}(5)) - 2 * \text{sqrt}(5) + 2) * \text{sqrt}(2)/16)$
$+ i * (((1+\text{sqrt}(5))*\text{sqrt}(10-2*\text{sqrt}(5)) + 2 * \text{sqrt}(5) - 2) * \text{sqrt}(2)/16)$

22: $((\text{sqrt}(15)+\text{sqrt}(3)-\text{sqrt}(10-2*\text{sqrt}(5)))/8) + i * ((\text{sqrt}(30-6*\text{sqrt}(5)) + \text{sqrt}(5) + 1)/8)$

23: $((2*(\text{sqrt}(15-3*\text{sqrt}(5)) + \text{sqrt}(5-\text{sqrt}(5))) - \text{sqrt}(30) + \text{sqrt}(10) - \text{sqrt}(6) + \text{sqrt}(2))/16)$
$+ i * ((\text{sqrt}(30)+\text{sqrt}(10)+\text{sqrt}(6)+\text{sqrt}(2)+(\text{sqrt}(6)-\text{sqrt}(2))*\text{sqrt}(10-2*\text{sqrt}(5)))/16)$

24: $((-1+\text{sqrt}(5))/4) + i * ((1+\text{sqrt}(5))*\text{sqrt}(10-2*\text{Sqrt}(5))/8)$

25: $((\text{sqrt}(6)-\text{sqrt}(2))/4) + i * ((\text{sqrt}(6)+\text{sqrt}(2))/4)$

26: $((2*\text{sqrt}(3)*(1-\text{sqrt}(5))+(1+\text{sqrt}(5))*\text{sqrt}(10-2*\text{sqrt}(5)))/16)$
$+ i * (((\text{sqrt}(15)+\text{sqrt}(3))*\text{sqrt}(10-2*\text{sqrt}(5)) + 2*\text{sqrt}(5)-2)/16)$

27: $((\text{SQRT}(10)+\text{SQRT}(2) -\text{SQRT}(20-4*\text{SQRT}(5)))/8)$
$+ i * (\text{sqrt}(2) * ((1+\text{sqrt}(5)) + \text{sqrt}(10 - 2 * \text{sqrt}(5))) /8)$

28: ((sqrt(30-6*sqrt(5)) - 1 -sqrt(5))/8) + i * ((sqrt(15)+sqrt(3)+sqrt(10-2*sqrt(5)))/8)

29: (((2*sqrt(30)+2*sqrt(10)-2*sqrt(6)-2*sqrt(2))
+ (sqrt(2)-sqrt(6)+sqrt(10)-sqrt(30))*sqrt(10-2*sqrt(5)))/32)
+ i * (((sqrt(30)+sqrt(2))*(sqrt(10-2*sqrt(5))+2)
+ (sqrt(10)+sqrt(6))*(sqrt(10-2*sqrt(5))-2))/32)

30: (0) + i * (1)

31: ((-((2*sqrt(30)+2*sqrt(10)-2*sqrt(6)-2*sqrt(2))
+ (sqrt(2)-sqrt(6)+sqrt(10)-sqrt(30))*sqrt(10-2*sqrt(5)))/32))
+ i * (((sqrt(30)+sqrt(2))*(sqrt(10-2*sqrt(5))+2) + (sqrt(10)+sqrt(6))*(sqrt(10-2*sqrt(5))-2))/32)

32: (-(sqrt(30-6*sqrt(5)) - 1 -sqrt(5))/8) + i * ((sqrt(15)+sqrt(3)+sqrt(10-2*sqrt(5)))/8)

33: ((-sqrt(10)-sqrt(2) +sqrt(20-4*sqrt(5)))/8)
+ i * (sqrt(2) * ((1+sqrt(5)) + sqrt(10 - 2 * sqrt(5))) /8)

34: ((2*sqrt(3)*(-1+sqrt(5))-(1+sqrt(5))*sqrt(10-2*sqrt(5)))/16)
+ i * (((sqrt(15)+sqrt(3))*sqrt(10-2*sqrt(5))   + 2*sqrt(5)-2)/16)

35: ((-sqrt(6)+sqrt(2))/4) + i * ((sqrt(6)+sqrt(2))/4)

36: ((1-sqrt(5))/4) + i * ((1+sqrt(5))*sqrt(10-2*Sqrt(5))/8)

37: ((-2*(sqrt(15-3*sqrt(5)) + sqrt(5-sqrt(5))) + sqrt(30) - sqrt(10) + sqrt(6) - sqrt(2))/16)
+ i * ((sqrt(30)+sqrt(10)+sqrt(6)+sqrt(2)+(sqrt(6)-sqrt(2))*sqrt(10-2*sqrt(5)))/16)

38: ((-sqrt(15)-sqrt(3)+sqrt(10-2*sqrt(5)))/8) + i * ((sqrt(30-6*sqrt(5)) + sqrt(5) + 1)/8)

39: ((-(1+sqrt(5))*sqrt(10-2*sqrt(5)) + 2 * sqrt(5) - 2) * sqrt(2)/16)
+ i * (((1+sqrt(5))*sqrt(10-2*sqrt(5)) + 2 * sqrt(5) - 2) * sqrt(2)/16)

40: (-1/2) + i * (sqrt(3)/2)

41: ((-(sqrt(30)-sqrt(10)+sqrt(6)-sqrt(2))*sqrt(10-2*sqrt(5))
-2*(sqrt(30)+sqrt(10)-sqrt(6)-sqrt(2)))/32)
+ i * (((sqrt(30)+sqrt(10)+sqrt(6)+sqrt(2))*sqrt(10-2*sqrt(5))
-2*(sqrt(30)-sqrt(10)-sqrt(6)+sqrt(2)))/32)

42: ((-sqrt(10-2*sqrt(5)))/4) + i * ((1+sqrt(5))/4)

43: ((-sqrt(2)) * (((1+sqrt(3))*(1 + sqrt(5)) +(1   - sqrt(3)) * sqrt(10-2*sqrt(5)))   ) /16)
+ i * ((sqrt(30)+sqrt(10)+sqrt(6)+sqrt(2)+(sqrt(6)-sqrt(2))*sqrt(10-2*sqrt(5)))/16
- (SQRT(10)+SQRT(2) -SQRT(20-4*SQRT(5)))/8)

44: (-(((sqrt(30)+sqrt(2))*(sqrt(10-2*sqrt(5))+2)
+ (sqrt(10)+sqrt(6))*(sqrt(10-2*sqrt(5))-2))/32
- ((2*sqrt(30)+2*sqrt(10)-2*sqrt(6)-2*sqrt(2))
+ (sqrt(2)-sqrt(6)+sqrt(10)-sqrt(30))*sqrt(10-2*sqrt(5)))/32) * (sqrt(2)/2))
+ i * (((((sqrt(30)+sqrt(2))*(sqrt(10-2*sqrt(5))+2) + (sqrt(10)+sqrt(6))*(sqrt(10-2*sqrt(5))-2))/32
+ ((2*sqrt(30)+2*sqrt(10)-2*sqrt(6)-2*sqrt(2))
+ (sqrt(2)-sqrt(6)+sqrt(10)-sqrt(30))*sqrt(10-2*sqrt(5)))/32) * (sqrt(2)/2))

45: (-sqrt(2)/2) + i * (sqrt(2)/2)

46: (-(((sqrt(30)+sqrt(2))*(sqrt(10-2*sqrt(5))+2)
+ (sqrt(10)+sqrt(6))*(sqrt(10-2*sqrt(5))-2))/32
+ ((2*sqrt(30)+2*sqrt(10)-2*sqrt(6)-2*sqrt(2))
+ (sqrt(2)-sqrt(6)+sqrt(10)-sqrt(30))*sqrt(10-2*sqrt(5)))/32) * (sqrt(2)/2))
+ i * (((((sqrt(30)+sqrt(2))*(sqrt(10-2*sqrt(5))+2) + (sqrt(10)+sqrt(6))*(sqrt(10-2*sqrt(5))-2))/32
- ((2*sqrt(30)+2*sqrt(10)-2*sqrt(6)-2*sqrt(2))
+ (sqrt(2)-sqrt(6)+sqrt(10)-sqrt(30))*sqrt(10-2*sqrt(5)))/32) * (sqrt(2)/2))

47: (-(sqrt(30)+sqrt(10)+sqrt(6)+sqrt(2)+(sqrt(6)-sqrt(2))*sqrt(10-2*sqrt(5)))/16
+ (SQRT(10)+SQRT(2) -SQRT(20-4*SQRT(5)))/8)
+ i * ((sqrt(2)) * (((1+sqrt(3))*(1 + sqrt(5)) +(1   - sqrt(3)) * sqrt(10-2*sqrt(5)))   ) /16)

48: (-(1+sqrt(5))/4) + i * (sqrt(10-2*sqrt(5))/4)

49: (-((sqrt(30)+sqrt(10)+sqrt(6)+sqrt(2))*sqrt(10-2*sqrt(5))
-2*(sqrt(30)-sqrt(10)-sqrt(6)+sqrt(2)))/32)
+ i * (((sqrt(30)-sqrt(10)+sqrt(6)-sqrt(2))*sqrt(10-2*sqrt(5))
+2*(sqrt(30)+sqrt(10)-sqrt(6)-sqrt(2)))/32)

50: (-sqrt(3)/2) + i * (1/2)

51: (-((1+sqrt(5))*sqrt(10-2*sqrt(5)) + 2 * sqrt(5) - 2) * sqrt(2)/16)
+ i * (((1+sqrt(5))*sqrt(10-2*sqrt(5)) - 2 * sqrt(5) + 2) * sqrt(2)/16)

52: (-(sqrt(30-6*sqrt(5)) + sqrt(5) + 1)/8) + i * ((sqrt(15)+sqrt(3)-sqrt(10-2*sqrt(5)))/8)

53: (-(sqrt(30)+sqrt(10)+sqrt(6)+sqrt(2)+(sqrt(6)-sqrt(2))*sqrt(10-2*sqrt(5)))/16)
+ i * ((2*(sqrt(15-3*sqrt(5)) + sqrt(5-sqrt(5))) - sqrt(30) + sqrt(10) - sqrt(6) + sqrt(2))/16)

54: (-(1+sqrt(5))*sqrt(10-2*Sqrt(5))/8) + i * ((-1+sqrt(5))/4)

55: (-(sqrt(6)+sqrt(2))/4) + i * ((sqrt(6)-sqrt(2))/4)

56: (-((sqrt(15)+sqrt(3))*sqrt(10-2*sqrt(5))   + 2*sqrt(5)-2)/16)
+ i * ((2*sqrt(3)*(1-sqrt(5))+(1+sqrt(5))*sqrt(10-2*sqrt(5)))/16)

57: (-sqrt(2) * ((1+sqrt(5)) + sqrt(10 - 2 * sqrt(5))) /8)
+ i * ((sqrt(10)+sqrt(2) -sqrt(20-4*sqrt(5)))/8)

58: (-(SQRT(15)+SQRT(3)+SQRT(10-2*SQRT(5)))/8)
+ i * ((sqrt(30-6*sqrt(5)) - 1 -sqrt(5))/8)

59: (-((sqrt(30)+sqrt(2))*(sqrt(10-2*sqrt(5))+2)
+ (sqrt(10)+sqrt(6))*(sqrt(10-2*sqrt(5))-2))/32)
+ i * (((2*sqrt(30)+2*sqrt(10)-2*sqrt(6)-2*sqrt(2))
+ (sqrt(2)-sqrt(6)+sqrt(10)-sqrt(30))*sqrt(10-2*sqrt(5)))/32)

60: (-1) + i * (0)

61: (-((sqrt(30)+sqrt(2))*(sqrt(10-2*sqrt(5))+2)
+ (sqrt(10)+sqrt(6))*(sqrt(10-2*sqrt(5))-2))/32)
+ i * (((-2*sqrt(30)-2*sqrt(10)+2*sqrt(6)+2*sqrt(2)) - (sqrt(2)-sqrt(6)+sqrt(10)-sqrt(30))
*sqrt(10-2*sqrt(5)))/32)

62: (-(SQRT(15)+SQRT(3)+SQRT(10-2*SQRT(5)))/8)
+ i * ((-sqrt(30-6*sqrt(5)) + 1 +sqrt(5))/8)

63: (-sqrt(2) * ((1+sqrt(5)) + sqrt(10 - 2 * sqrt(5))) /8)
+ i * ((-sqrt(10)-sqrt(2) +sqrt(20-4*sqrt(5)))/8)

64: (-((sqrt(15)+sqrt(3))*sqrt(10-2*sqrt(5))   + 2*sqrt(5)-2)/16)
+ i * ((-2*sqrt(3)*(1-sqrt(5))-(1+sqrt(5))*sqrt(10-2*sqrt(5)))/16)

65: (-(sqrt(6)+sqrt(2))/4) + i * ((-sqrt(6)+sqrt(2))/4)

66: ((-(1+sqrt(5)))*(sqrt(10-2*sqrt(5)))/8) + i * ((1-sqrt(5))/4)

67: (-(sqrt(30)+sqrt(10)+sqrt(6)+sqrt(2)+(sqrt(6)-sqrt(2))*sqrt(10-2*sqrt(5)))/16)
+ i * ((-2*(sqrt(15-3*sqrt(5)) + sqrt(5-sqrt(5))) + sqrt(30) - sqrt(10) + sqrt(6) - sqrt(2))/16)

68: (-(sqrt(30-6*sqrt(5)) + sqrt(5) + 1)/8) + i * ((-sqrt(15)-sqrt(3)+sqrt(10-2*sqrt(5)))/8)

69: (-((1+sqrt(5))*sqrt(10-2*sqrt(5)) + 2 * sqrt(5) - 2) * sqrt(2)/16)
+ i * ((-(1+sqrt(5))*sqrt(10-2*sqrt(5)) + 2 * sqrt(5) - 2) * sqrt(2)/16)

70: (-sqrt(3)/2) + i * (-1/2)

71: (-((sqrt(30)+sqrt(10)+sqrt(6)+sqrt(2))*sqrt(10-2*sqrt(5))
-2*(sqrt(30)-sqrt(10)-sqrt(6)+sqrt(2)))/32)
+ i * ((-(sqrt(30)-sqrt(10)+sqrt(6)-sqrt(2))*sqrt(10-2*sqrt(5))
-2*(sqrt(30)+sqrt(10)-sqrt(6)-sqrt(2)))/32)

72: (-(1+sqrt(5))/4) + i * ((-sqrt(10-2*sqrt(5)))/4)

73: (-(sqrt(30)+sqrt(10)+sqrt(6)+sqrt(2)+(sqrt(6)-sqrt(2))*sqrt(10-2*sqrt(5)))/16
+ (SQRT(10)+SQRT(2) -SQRT(20-4*SQRT(5)))/8)
+ i * ((-sqrt(2)) * (((1+sqrt(3))*(1 + sqrt(5)) +(1   - sqrt(3)) * sqrt(10-2*sqrt(5)))   ) /16)

74: (-(((sqrt(30)+sqrt(2))*(sqrt(10-2*sqrt(5))+2)
+ (sqrt(10)+sqrt(6))*(sqrt(10-2*sqrt(5))-2))/32
+ ((2*sqrt(30)+2*sqrt(10)-2*sqrt(6)-2*sqrt(2))
+ (sqrt(2)-sqrt(6)+sqrt(10)-sqrt(30))*sqrt(10-2*sqrt(5)))/32) * (sqrt(2)/2))
+ i * (-(((sqrt(30)+sqrt(2))*(sqrt(10-2*sqrt(5))+2) + (sqrt(10)+sqrt(6))*(sqrt(10-2*sqrt(5))-2))/32
- ((2*sqrt(30)+2*sqrt(10)-2*sqrt(6)-2*sqrt(2))
+ (sqrt(2)-sqrt(6)+sqrt(10)-sqrt(30))*sqrt(10-2*sqrt(5)))/32) * (sqrt(2)/2))

75: (-sqrt(2)/2) + i * (-sqrt(2)/2)

76: (-(((sqrt(30)+sqrt(2))*(sqrt(10-2*sqrt(5))+2)
+ (sqrt(10)+sqrt(6))*(sqrt(10-2*sqrt(5))-2))/32 - ((2*sqrt(30)+2*sqrt(10)-2*sqrt(6)-2*sqrt(2))
+ (sqrt(2)-sqrt(6)+sqrt(10)-sqrt(30))*sqrt(10-2*sqrt(5)))/32) * (sqrt(2)/2))
+ i * (-(((sqrt(30)+sqrt(2))*(sqrt(10-2*sqrt(5))+2) + (sqrt(10)+sqrt(6))*(sqrt(10-2*sqrt(5))-2))/32
+ ((2*sqrt(30)+2*sqrt(10)-2*sqrt(6)-2*sqrt(2)) + (sqrt(2)-sqrt(6)+sqrt(10)-sqrt(30))
*sqrt(10-2*sqrt(5)))/32) * (sqrt(2)/2))

77: ((-sqrt(2)) * (((1+sqrt(3))*(1 + sqrt(5)) +(1   - sqrt(3)) * sqrt(10-2*sqrt(5)))   ) /16)
+ i * (-(sqrt(30)+sqrt(10)+sqrt(6)+sqrt(2)+(sqrt(6)-sqrt(2))*sqrt(10-2*sqrt(5)))/16
+ (SQRT(10)+SQRT(2) -SQRT(20-4*SQRT(5)))/8)

78: ((-sqrt(10-2*sqrt(5)))/4) + i * (-(1+sqrt(5))/4)

79: ((-(sqrt(30)-sqrt(10)+sqrt(6)-sqrt(2))*sqrt(10-2*sqrt(5))
-2*(sqrt(30)+sqrt(10)-sqrt(6)-sqrt(2)))/32)
+ i * (-((sqrt(30)+sqrt(10)+sqrt(6)+sqrt(2))*sqrt(10-2*sqrt(5))
-2*(sqrt(30)-sqrt(10)-sqrt(6)+sqrt(2)))/32)

80: (-1/2) + i * ((-sqrt(3)/2))

81: ((-(1+sqrt(5))*sqrt(10-2*sqrt(5)) + 2 * sqrt(5) - 2) * sqrt(2)/16)
+ i * (-((1+sqrt(5))*sqrt(10-2*sqrt(5)) + 2 * sqrt(5) - 2) * sqrt(2)/16)

82: ((-sqrt(15)-sqrt(3)+sqrt(10-2*sqrt(5)))/8) + i * (-(sqrt(30-6*sqrt(5)) + sqrt(5) + 1)/8)

83: ((-2*(sqrt(15-3*sqrt(5)) + sqrt(5-sqrt(5))) + sqrt(30) - sqrt(10) + sqrt(6) - sqrt(2))/16)
+ i * (-(sqrt(30)+sqrt(10)+sqrt(6)+sqrt(2)+(sqrt(6)-sqrt(2))*sqrt(10-2*sqrt(5)))/16)

84: ((1-sqrt(5))/4) + i * (-(1+sqrt(5))*sqrt(10-2*Sqrt(5))/8)

85: ((-sqrt(6)+sqrt(2))/4) + i * (-(sqrt(6)+sqrt(2))/4)

86: ((-2*sqrt(3)*(1-sqrt(5))-(1+sqrt(5))*sqrt(10-2*sqrt(5)))/16)
+ i * (-((sqrt(15)+sqrt(3))*sqrt(10-2*sqrt(5)) + 2*sqrt(5)-2)/16)

87: ((-sqrt(10)-sqrt(2) +sqrt(20-4*sqrt(5)))/8)
+ i * (-sqrt(2) * ((1+sqrt(5)) + sqrt(10 - 2 * sqrt(5))) /8)

88: ((-sqrt(30-6*sqrt(5)) + 1 +sqrt(5))/8)
+ i * (-(SQRT(15)+SQRT(3)+SQRT(10-2*SQRT(5)))/8)

89: (((-2*sqrt(30)-2*sqrt(10)+2*sqrt(6)+2*sqrt(2))
- (sqrt(2)-sqrt(6)+sqrt(10)-sqrt(30))*sqrt(10-2*sqrt(5)))/32)
+ i * (-((sqrt(30)+sqrt(2))*(sqrt(10-2*sqrt(5))+2) + (sqrt(10)+sqrt(6))*(sqrt(10-2*sqrt(5))-2))/32)
90: (0) + i * (-1)

91: (((2*sqrt(30)+2*sqrt(10)-2*sqrt(6)-2*sqrt(2))
+ (sqrt(2)-sqrt(6)+sqrt(10)-sqrt(30))*sqrt(10-2*sqrt(5)))/32)
+ i * (-((sqrt(30)+sqrt(2))*(sqrt(10-2*sqrt(5))+2) + (sqrt(10)+sqrt(6))*(sqrt(10-2*sqrt(5))-2))/32)

92: ((sqrt(30-6*sqrt(5)) - 1 -sqrt(5))/8)
+ i * (-(SQRT(15)+SQRT(3)+SQRT(10-2*SQRT(5)))/8)

93: ((sqrt(10)+sqrt(2) -sqrt(20-4*sqrt(5)))/8)
+ i * (-sqrt(2) * ((1+sqrt(5)) + sqrt(10 - 2 * sqrt(5))) /8)

94: ((2*sqrt(3)*(1-sqrt(5))+(1+sqrt(5))*sqrt(10-2*sqrt(5)))/16)
+ i * (-((sqrt(15)+sqrt(3))*sqrt(10-2*sqrt(5)) + 2*sqrt(5)-2)/16)

95: ((sqrt(6)-sqrt(2))/4) + i * (-(sqrt(6)+sqrt(2))/4)

96: ((-1+sqrt(5))/4) + i * ((-(1+sqrt(5)))*(sqrt(10-2*sqrt(5)))/8)

97: ((2*(sqrt(15-3*sqrt(5)) + sqrt(5-sqrt(5))) - sqrt(30) + sqrt(10) - sqrt(6) + sqrt(2))/16)
+ i * (-(sqrt(30)+sqrt(10)+sqrt(6)+sqrt(2)+(sqrt(6)-sqrt(2))*sqrt(10-2*sqrt(5)))/16)

98: ((sqrt(15)+sqrt(3)-sqrt(10-2*sqrt(5)))/8) + i * (-(sqrt(30-6*sqrt(5)) + sqrt(5) + 1)/8)

99: ((((1+sqrt(5))*sqrt(10-2*sqrt(5)) - 2 * sqrt(5) + 2) * sqrt(2)/16)
+ i * (-((1+sqrt(5))*sqrt(10-2*sqrt(5)) + 2 * sqrt(5) - 2) * sqrt(2)/16)

100: (1/2) + i * ((-sqrt(3)/2))

101: ((((sqrt(30)-sqrt(10)+sqrt(6)-sqrt(2))*sqrt(10-2*sqrt(5))
+2*(sqrt(30)+sqrt(10)-sqrt(6)-sqrt(2)))/32)
+ i * (-((sqrt(30)+sqrt(10)+sqrt(6)+sqrt(2))*sqrt(10-2*sqrt(5))
-2*(sqrt(30)-sqrt(10)-sqrt(6)+sqrt(2)))/32)

102: (sqrt(10-2*sqrt(5))/4) + i * (-(1+sqrt(5))/4)

103: ((sqrt(2)) * (((1+sqrt(3))*(1 + sqrt(5)) +(1   - sqrt(3)) * sqrt(10-2*sqrt(5)))   ) /16)
+ i * (-(sqrt(30)+sqrt(10)+sqrt(6)+sqrt(2)+(sqrt(6)-sqrt(2))*sqrt(10-2*sqrt(5)))/16
+ (SQRT(10)+SQRT(2) -SQRT(20-4*SQRT(5)))/8)

104: (((((sqrt(30)+sqrt(2))*(sqrt(10-2*sqrt(5))+2)
+ (sqrt(10)+sqrt(6))*(sqrt(10-2*sqrt(5))-2))/32
- ((2*sqrt(30)+2*sqrt(10)-2*sqrt(6)-2*sqrt(2))
+ (sqrt(2)-sqrt(6)+sqrt(10)-sqrt(30))*sqrt(10-2*sqrt(5)))/32) * (sqrt(2)/2))
+ i * (-(((sqrt(30)+sqrt(2))*(sqrt(10-2*sqrt(5))+2) + (sqrt(10)+sqrt(6))*(sqrt(10-2*sqrt(5))-2))/32
+ ((2*sqrt(30)+2*sqrt(10)-2*sqrt(6)-2*sqrt(2))
+ (sqrt(2)-sqrt(6)+sqrt(10)-sqrt(30))*sqrt(10-2*sqrt(5)))/32) * (sqrt(2)/2))

105: (sqrt(2)/2) + i * (-sqrt(2)/2)

106: (((((sqrt(30)+sqrt(2))*(sqrt(10-2*sqrt(5))+2)
+ (sqrt(10)+sqrt(6))*(sqrt(10-2*sqrt(5))-2))/32 + ((2*sqrt(30)+2*sqrt(10)-2*sqrt(6)-2*sqrt(2))
+ (sqrt(2)-sqrt(6)+sqrt(10)-sqrt(30))*sqrt(10-2*sqrt(5)))/32) * (sqrt(2)/2))
+ i * (-(((sqrt(30)+sqrt(2))*(sqrt(10-2*sqrt(5))+2) + (sqrt(10)+sqrt(6))*(sqrt(10-2*sqrt(5))-2))/32
- ((2*sqrt(30)+2*sqrt(10)-2*sqrt(6)-2*sqrt(2))
+ (sqrt(2)-sqrt(6)+sqrt(10)-sqrt(30))*sqrt(10-2*sqrt(5)))/32) * (sqrt(2)/2))

107: ((sqrt(30)+sqrt(10)+sqrt(6)+sqrt(2)+(sqrt(6)-sqrt(2))*sqrt(10-2*sqrt(5)))/16
- (SQRT(10)+SQRT(2) -SQRT(20-4*SQRT(5)))/8)
+ i * ((-sqrt(2)) * (((1+sqrt(3))*(1 + sqrt(5)) +(1   - sqrt(3)) * sqrt(10-2*sqrt(5)))   ) /16)

108: ((1+sqrt(5))/4) + i * ((-sqrt(10-2*sqrt(5)))/4)

109: ((((sqrt(30)+sqrt(10)+sqrt(6)+sqrt(2))*sqrt(10-2*sqrt(5))
-2*(sqrt(30)-sqrt(10)-sqrt(6)+sqrt(2)))/32)
+ i * ((-(sqrt(30)-sqrt(10)+sqrt(6)-sqrt(2))*sqrt(10-2*sqrt(5))
-2*(sqrt(30)+sqrt(10)-sqrt(6)-sqrt(2)))/32)

110: (Sqrt(3)/2) + i * (-1/2)

111: ((((1+sqrt(5))*sqrt(10-2*sqrt(5)) + 2 * sqrt(5) - 2) * sqrt(2)/16)
+ i * ((-(1+sqrt(5))*sqrt(10-2*sqrt(5)) + 2 * sqrt(5) - 2) * sqrt(2)/16)

112: ((sqrt(30-6*sqrt(5)) + sqrt(5) + 1)/8) + i * ((-sqrt(15)-sqrt(3)+sqrt(10-2*sqrt(5)))/8)

113: ((sqrt(30)+sqrt(10)+sqrt(6)+sqrt(2)+(sqrt(6)-sqrt(2))*sqrt(10-2*sqrt(5)))/16)
+ i * ((-2*(sqrt(15-3*sqrt(5)) + sqrt(5-sqrt(5))) + sqrt(30) - sqrt(10) + sqrt(6) - sqrt(2))/16)

114: ((1+sqrt(5))*sqrt(10-2*Sqrt(5))/8) + i * ((1-sqrt(5))/4)

115: ((sqrt(6)+sqrt(2))/4) + i * ((-sqrt(6)+sqrt(2))/4)

116: (((sqrt(15)+sqrt(3))*sqrt(10-2*sqrt(5)) + 2*sqrt(5)-2)/16)
+ i * ((-2*sqrt(3)*(1-sqrt(5))-(1+sqrt(5))*sqrt(10-2*sqrt(5)))/16)

117: (sqrt(2) * ((1+sqrt(5)) + sqrt(10 - 2 * sqrt(5))) /8)
+ i * ((-sqrt(10)-sqrt(2) +sqrt(20-4*sqrt(5)))/8)

118: ((sqrt(15)+sqrt(3)+sqrt(10-2*sqrt(5)))/8) + i * ((-sqrt(30-6*sqrt(5)) + 1 +sqrt(5))/8)

119: (((sqrt(30)+sqrt(2))*(sqrt(10-2*sqrt(5))+2)
+ (sqrt(10)+sqrt(6))*(sqrt(10-2*sqrt(5))-2))/32)
+ i * (((-2*sqrt(30)-2*sqrt(10)+2*sqrt(6)+2*sqrt(2))
- (sqrt(2)-sqrt(6)+sqrt(10)-sqrt(30))*sqrt(10-2*sqrt(5)))/32)

120: (1) + i * (0)

Nth derivative of a product

Notation:

The nth power of y is written y^n.

The nth derivative of y is written y@n.

The first derivative of y is written y @ 1 or y'
The second derivation of y is written y @ 2 or y''
etc

Note the following parallel between ^ and @

(x + y) ^ n
(u v) @ n

(x + y) ^ 0 = x^0 y^0 = 1
(u v) @ 0 =(u @ 0) (v @ 0) = u v

(x + y) ^1 = x^1 y^0 + y^1 x^0
(u v) @ 1 = (u @ 1)(v @ 0) + (u @ 0) (v @ 1)

x + y
 u' v + u v'

(x + y) ^ 2 = x^2 y^0 + 2 x^1 y^1 + x^0 y^2= x^2 + 2 x y + y^2

(u v)@ 2 = (u @ 2) (v @ 0) + 2 (u @ 1) (v @ 1) + (u @ 0) (v @ 2)= u'' v + 2 u' v' + u v''

(x + y) ^3 = x^3 y^0 + 3 x^2 y^1 + 3 x^1 y^2 + x^0 y^3

(u v) @ 3
= (u@3) (v@0) + 3 (u@2) (v@1) + 3 (u@1) (v@2) + (u@0) (v@3)
= u''' v + 3 u'' v' + 3 u' v'' + u v'''

etc

Now, what does the formula for (x + y)^n give if we put n = -1?

(x + y) ^ (-1)

= x^(-1) y^0 - x^(-2) y^1 + x^(-3) y^2 - x^(-4) y^3 + ...

So the integral of the product of two functions can be given the formula

(u v) @ (-1) = u@(-1) v - u@(-2) v' + u@(-3) v" - u@(-4) v''' + ...

This makes it easy to integrate the product of an exponential function with a polynomial. We can find any order integral of the exponential function easily, and the successive derivatives of the polynomial eventually reach 0. Therefore, we would identify the exponential function with u and the polynomial function with v in the above formula.

Suppose we wanted to find the integral of x sin(x) ?

(u v) @ (-1) = u@(-1) v - u@(-2) v' + u@(-3) v" - u@(-4) v''' + ...

Set u = sin(x)   and v = x

u@(-1) = - cos(x)   v' = 1

u@(-2) = - sin(x)   v" = 0

(x sin(x)) @ (-1) = [- cos(x)] x - [- sin(x)] [1] + 0
         = - x cos(x) + sin(x)

And you can verify by taking derivative.

w = - x cos(x) + sin(x)

w' = - x [- sin(x)] - cos(x) + cos(x)
   = x sin(x)

Fractional Derivatives

In calculus, the first derivation of x**n is n*x**(n-1).
Applying the same rule: Multiply the coefficient by the exponent, and reduce the exponent by 1, gives second derivative of
n*(n-1) * x**(n-2).

Third derivative is n*(n-1)*(n-2) * x**(n-3).

How would we write the kth derivative for some positive integer, k?

kth derivative is:

n*(n-1)*(n-2)*...(n-k+1) * x**(n-k)

We can write this more compactly as:

kth derivative is:

(n! / (n-k)!) * x**(n-k).

Now that we have this compact form, we can consider what it would mean if k were not an integer.

Following this form, we can say that the (1/2) derivative is (n!/(n-1/2)!) * x**(n-1/2)

and the (1/2) derivative of the (1/2) derivative is:

(n!/(n-1/2)!) * ((n-1/2)! / (n-1)!) * x**(n-1/2 - 1/2)

= (n! / (n-1)!) * x**(n-1) which is consistent with the first derivative.

Is it possible to assign a meaning to ((n-1/2)!)?

Yes. The gamma function extends the factorial function to non-integers.

http://mathworld.wolfram.com/GammaFunction.html

http://www.sosmath.com/calculus/improper/gamma/gamma.html

For integer values of k, gamma(k+1) = k!.

kth derivative of x**n is:

(n! / (n-k)!) * x**(n-k)
= (gamma(n+1)/gamma(n-k+1)) * x**(n-k)

(1/2) derivative of x**n is:

(gamma(n+1)/gamma(n+1/2)) * x**(n-1/2).

It happens that gamma(n+1/2) can be calculated exactly.

Note that gamma(n+1) = n! = n*(n-1)*(n-2)*...2*1

(n+1) gamma(n+1) = (n+1)*n! = (n+1)!

(n+1) gamma(n+1) = gamma(n+2)

In general, for any number z, for which gamma(z) is defined,
z*gamma(z) = gamma(z+1).

Thus gamma(1/2 + 1) = (1/2) gamma(1/2)

gamma(3/2) = (1/2) gamma(1/2)
gamma(5/2) = (3/2)(1/2) gamma(1/2)
gamma(7/2) = (7/2)(3/2)(1/2) gamma (1/2)

Thus we can calculate gamma(n+1/2), for any not too large integer n, if we can calculate gamma(1/2).

It is, in fact easy to calculate gamma(1/2).

gamma(1/2) = sqrt(pi).

Sum Like Powers

You have probably seen the formula for the sum of the first n positive integers, and maybe even the formula for the sum of the first n square integers and first n cube integers.

There is a simple rule for progressing from a formula for the sum of the first n integers raised to a power to the formula for the sum of the first n integers raised to the next highest power.

To illustrate this rule, start with the zeroth power.

$s_0(n) = 1^0 + 2^0 + 3^0 + ... + n^0$.

It is easily seen that the formula for this sum is
$s_0(n) = n$.

We next illustrate the rule by applying it to s0(n) to get the formula for
$s_1(n) = 1 + 2 + 3 + 4 + ... + n$.

Multiply $s_0(n)$ by the next higher power, namely 1.

$1 * s_0(n) = n$.

Now we apply the calculus rule for integration.

For each term in the polynomial $1 * s_0(n) = n$, increase the exponent by 1, and divide the coefficient by the new exponent.

$n \longrightarrow n^2/2$

$s_1(n) = n^2/2 +$ some unknown coefficient $* n$.

How do we determine that unknown coefficient?

We know that $s_1(1) = 1$ by the definition of $s_1(1) = 1^1$.

$s_1(n) = n^2/2 + (1 - 1/2)* n = (1/2) n^2 + (1/2) n$.

You probably recognize this formula as equivalent to
$s_1(n) = (1/2)n(n+1)$.

For the $s_2(n) = 1^2 + 2^2 + 3^2 + ... + n^2$
we start with

$s_1(n) = (1/2) n^2 + (1/2) n$

Multiply by 2, the next power up.

$2 * s_1(n) = n^2 + n$

Do the calculus integration rule.

$n^2 \longrightarrow (1/3) n^3$
$n \longrightarrow (1/2) n^2$
$s_2(n) = (1/3) n^3 + (1/2) n^2 + (1 - (1/3) - (1/2)) n$
$s_2(n) = (1/3) n^3 + (1/2) n^2 + (1/6) n$

Next, we find the formula for

$s_3(n) = 1^3 + 2^3 + 3^3 + 4^3 + ... + n^3$

Start with

$s_2(n) = (1/3) n^3 + (1/2) n^2 + (1/6) n$

Multiply by 3, the new exponent.

$3 * s_2(n) = n^3 + (3/2) n^2 + (3/6) n$

Apply the integration rule.

$n^3 \longrightarrow (1/4) n^4$
$(3/2) n^2 \longrightarrow (1/2) n^3$
$(3/6) n \longrightarrow (1/4) n^2$

$s_3(n) = (1/4) n^4 + (1/2) n^3 + (1/4) n^2 + (1 - (1/4) - (1/2) - (1/4)) n$

$s3(n) = (1/4) n^4 + (1/2) n^3 + (1/4) n^2$
because
$(1 - (1/4) - (1/2) - (1/4)) = 0.$

$s_1(n) = (1/2) n^2 + (1/2) n$

$s_2(n) = (1/3) n^3 + (1/2) n^2 + (1/6)n$

$s_3(n) = (1/4) n^4 + (1/2) n^3 + (1/4) n^2$

$s_4(n) = (1/5) n^5 + (1/2) n^4 + (1/3) n^3 - (1/30) n$

$s_5(n) = (1/6) n^6 + (1/2) n^5 + (5/12) n^4 - (1/12) n^2$

$s_6(n) = (1/7) n^7 + (1/2) n^6 + (1/2) n^5 - (1/6) n^3 + (1/42) n$

$s_7(n) = (1/8) n^8 + (1/2) n^7 + (7/12) n^6 - (7/24) n^4 + (1/12) n^2$

$s_8(n) = (1/9) n^9 + (1/2) n^8 + (2/3) n^7 - (7/15) n^5 + (2/9) n^3 - (1/30) n$

$s_9(n) = (1/10) n^{10} + (1/2) n^9 + (3/4) n^8 - (7/10) n^6 + (1/2) n^4 - (3/20) n^2$

$s_{10}(n) = (1/11) n^{11} + (1/2) n^{10} + (5/6) n^9 - n^7 + n^5 - (1/2) n^3 + (5/66) n$

$s_{11}(n) = (1/12) n^{12} + (1/2) n^{11} + (11/12) n^{10} - (11/8) n^8 + (11/6) n^6 - (11/8) n^4$
$+ (5/12) n^2$

$s_{12}(n) = (1/13) n^{13} + (1/2) n^{12} + n^{11} - (11/6) n^9 + (22/7) n^7 - (33/10) n^5 + (5/3) n^3$
$- (691/ 2730) n$

$s_{13}(n) = (1/14) n^{14} + (1/2) n^{13} + (13/12) n^{12} - (143/60) n^{10} + (143/28) n^8 - (143/20) n^6$
$+ (65/12) n^4 - (691/420) n^2$

$s_{14}(n) = (1/15) n^{15} + (1/2) n^{14} + (7/6) n^{13} - (91/30) n^{11} + (143/18) n^9 - (143/10) n^7$
$+ (91/6) n^5 - (691/90) n^3 + (7/6) n$

$s_{15}(n) = (1/16) n^{16} + (1/2) n^{15} + (5/4) n^{14} - (91/24) n^{12} + (143/12) n^{10} - (429/16) n^8$
$+ (455/12) n^6 - (691/24) n^4 + (35/4) n^2$

$s_{16}(n) = (1/17) n^{17} + (1/2) n^{16} + (4/3) n^{15} - (14/3) n^{13} + (52/3) n^{11} - (143/3) n^9$
$+ (260/3) n^7 - (1382/15) n^5 + (140/3) n^3 - (3617/510) n$

$s_{17}(n) = (1/18) n^{18} + (1/2) n^{17} + (17/12) n^{16} - (17/3) n^{14} + (221/9) n^{12} - (2431/30) n^{10}$
$+ (1105/6) n^8 - (11747/45) n^6 + (595/3) n^4 - (3617/60) n^2$

$s_{18}(n) = (1/19) n^{19} + (1/2) n^{18} + (3/2) n^{17} - (34/5) n^{15} + 34 n^{13} - (663/5) n^{11}$
$+ (1105/3) n^9 - (23494/35) n^7 + 714 n^5 - (3617/10) n^3 + (43867/798) n$

$s_{19}(n) = (1/20) n^{20} + (1/2) n^{19} + (19/12) n^{18} - (323/40) n^{16} + (323/7) n^{14}$
$- (4199/20) n^{12} + (4199/6) n^{10} - (223193/140) n^8 + 2261 n^6 - (68723/40) n^4$
$+ (43867/84) n^2$

$s_{20}(n) = (1/21) n^{21} + (1/2) n^{20} + (5/3) n^{19} - (19/2) n^{17} + (1292/21) n^{15} - 323 n^{13}$
$+ (41990/33) n^{11} - (223193/63) n^9 + 6460 n^7 - (68723/10) n^5 + (219335/63) n^3$
$- (174611/330) n$

$s_{21}(n) = (1/22) n^{22} + (1/2) n^{21} + (7/4) n^{20} - (133/12) n^{18} + (323/4) n^{16} - (969/2) n^{14}$
$+ (146965/66) n^{12} - (223193/30) n^{10} + (33915/2) n^8 - (481061/20) n^6$
$+ (219335/12) n^4 - (1222277/ 220) n^2$

$s_{22}(n) = (1/23) n^{23} + (1/2) n^{22} + (11/6) n^{21} - (77/6) n^{19} + (209/2) n^{17} - (3553/5) n^{15}$
$+ (11305/3) n^{13} - (223193/15) n^{11} + (124355/3) n^9 - (755953/10) n^7$
$+ (482537/6) n^5 - (1222277/30) n^3 + (854513/138) n$

$s_{23}(n) = (1/24)\,n^{24} + (1/2)\,n^{23} + (23/12)\,n^{22} - (1771/120)\,n^{20} + (4807/36)\,n^{18}$
$\qquad - (81719/80)\,n^{16} + (37145/6)\,n^{14} - (5133439/180)\,n^{12} + (572033/6)\,n^{10}$
$\qquad - (17386919/80)\,n^{8} + (11098351/36)\,n^{6} - (28112371/120)\,n^{4} + (854513/12)\,n^{2}$

$s_{24}(n) = (1/25)\,n^{25} + (1/2)\,n^{24} + 2\,n^{23} - (253/15)\,n^{21} + (506/3)\,n^{19} - (14421/10)\,n^{17}$
$\qquad + (29716/3)\,n^{15} - (10266878/195)\,n^{13} + 208012\,n^{11} - (17386919/30)\,n^{9}$
$\qquad + (22196702/21)\,n^{7} - (28112371/25)\,n^{5} + (1709026/3)\,n^{3} - (236364091/2730)\,n$

$s_{25} = (1/26)\,n^{26} + (1/2)\,n^{25} + (25/12)\,n^{24} - (115/6)\,n^{22} + (1265/6)\,n^{20}$
$\qquad - (24035/12)\,n^{18} + (185725/12)\,n^{16} - (25667195/273)\,n^{14} + (1300075/3)\,n^{12}$
$\qquad - (17386919/12)\,n^{10} + (277458775/84)\,n^{8} - (28112371/6)\,n^{6} + (21362825/6)\,n^{4}$
$\qquad - (1181820455/1092)\,n^{2}$

Change for a Dollar

Occasionally one sees posted on the internet that there are 293 ways to change a dollar. This is not quite true. There are 292 ways to make change for a dollar. The 293rd way was supposedly to change a dollar into a dollar, which I do not count as a valid change for a dollar.

N	Penny 1	Nickel 5	Dime 10	Quarter 25	Half Dollar 50
1	100	0	0	0	0
2	95	1	0	0	0
3	90	2	0	0	0
4	85	3	0	0	0
5	80	4	0	0	0
6	75	5	0	0	0
7	70	6	0	0	0
8	65	7	0	0	0
9	60	8	0	0	0
10	55	9	0	0	0
11	50	10	0	0	0
12	45	11	0	0	0
13	40	12	0	0	0
14	35	13	0	0	0
15	30	14	0	0	0
16	25	15	0	0	0
17	20	16	0	0	0
18	15	17	0	0	0
19	10	18	0	0	0
20	5	19	0	0	0
21	0	20	0	0	0
22	90	0	1	0	0
23	85	1	1	0	0
24	80	2	1	0	0
25	75	3	1	0	0
26	70	4	1	0	0
27	65	5	1	0	0
28	60	6	1	0	0
29	55	7	1	0	0
30	50	8	1	0	0
31	45	9	1	0	0
32	40	10	1	0	0
33	35	11	1	0	0

	Penny	Nickel	Dime	Quarter	Half Dollar
N	1	5	10	25	50
34	30	12	1	0	0
35	25	13	1	0	0
36	20	14	1	0	0
37	15	15	1	0	0
38	10	16	1	0	0
39	5	17	1	0	0
40	0	18	1	0	0
41	80	0	2	0	0
42	75	1	2	0	0
43	70	2	2	0	0
44	65	3	2	0	0
45	60	4	2	0	0
46	55	5	2	0	0
47	50	6	2	0	0
48	45	7	2	0	0
49	40	8	2	0	0
50	35	9	2	0	0
51	30	10	2	0	0
52	25	11	2	0	0
53	20	12	2	0	0
54	15	13	2	0	0
55	10	14	2	0	0
56	5	15	2	0	0
57	0	16	2	0	0
58	70	0	3	0	0
59	65	1	3	0	0
60	60	2	3	0	0
61	55	3	3	0	0
62	50	4	3	0	0
63	45	5	3	0	0
64	40	6	3	0	0
65	35	7	3	0	0
66	30	8	3	0	0
67	25	9	3	0	0
68	20	10	3	0	0
69	15	11	3	0	0
70	10	12	3	0	0
71	5	13	3	0	0

N	Penny 1	Nickel 5	Dime 10	Quarter 25	Half Dollar 50
72	0	14	3	0	0
73	60	0	4	0	0
74	55	1	4	0	0
75	50	2	4	0	0
76	45	3	4	0	0
77	40	4	4	0	0
78	35	5	4	0	0
79	30	6	4	0	0
80	25	7	4	0	0
81	20	8	4	0	0
82	15	9	4	0	0
83	10	10	4	0	0
84	5	11	4	0	0
85	0	12	4	0	0
86	50	0	5	0	0
87	45	1	5	0	0
88	40	2	5	0	0
89	35	3	5	0	0
90	30	4	5	0	0
91	25	5	5	0	0
92	20	6	5	0	0
93	15	7	5	0	0
94	10	8	5	0	0
95	5	9	5	0	0
96	0	10	5	0	0
97	40	0	6	0	0
98	35	1	6	0	0
99	30	2	6	0	0
100	25	3	6	0	0
101	20	4	6	0	0
102	15	5	6	0	0
103	10	6	6	0	0
104	5	7	6	0	0
105	0	8	6	0	0
106	30	0	7	0	0
107	25	1	7	0	0
108	20	2	7	0	0
109	15	3	7	0	0

N	Penny 1	Nickel 5	Dime 10	Quarter 25	Half Dollar 50
110	10	4	7	0	0
111	5	5	7	0	0
112	0	6	7	0	0
113	20	0	8	0	0
114	15	1	8	0	0
115	10	2	8	0	0
116	5	3	8	0	0
117	0	4	8	0	0
118	10	0	9	0	0
119	5	1	9	0	0
120	0	2	9	0	0
121	0	0	10	0	0
122	75	0	0	1	0
123	70	1	0	1	0
124	65	2	0	1	0
125	60	3	0	1	0
126	55	4	0	1	0
127	50	5	0	1	0
128	45	6	0	1	0
129	40	7	0	1	0
130	35	8	0	1	0
131	30	9	0	1	0
132	25	10	0	1	0
133	20	11	0	1	0
134	15	12	0	1	0
135	10	13	0	1	0
136	5	14	0	1	0
137	0	15	0	1	0
138	65	0	1	1	0
139	60	1	1	1	0
140	55	2	1	1	0
141	50	3	1	1	0
142	45	4	1	1	0
143	40	5	1	1	0
144	35	6	1	1	0
145	30	7	1	1	0
146	25	8	1	1	0
147	20	9	1	1	0

	Penny	Nickel	Dime	Quarter	Half Dollar
N	1	5	10	25	50
148	15	10	1	1	0
149	10	11	1	1	0
150	5	12	1	1	0
151	0	13	1	1	0
152	55	0	2	1	0
153	50	1	2	1	0
154	45	2	2	1	0
155	40	3	2	1	0
156	35	4	2	1	0
157	30	5	2	1	0
158	25	6	2	1	0
159	20	7	2	1	0
160	15	8	2	1	0
161	10	9	2	1	0
162	5	10	2	1	0
163	0	11	2	1	0
164	45	0	3	1	0
165	40	1	3	1	0
166	35	2	3	1	0
167	30	3	3	1	0
168	25	4	3	1	0
169	20	5	3	1	0
170	15	6	3	1	0
171	10	7	3	1	0
172	5	8	3	1	0
173	0	9	3	1	0
174	35	0	4	1	0
175	30	1	4	1	0
176	25	2	4	1	0
177	20	3	4	1	0
178	15	4	4	1	0
179	10	5	4	1	0
180	5	6	4	1	0
181	0	7	4	1	0
182	25	0	5	1	0
183	20	1	5	1	0
184	15	2	5	1	0
185	10	3	5	1	0

N	Penny 1	Nickel 5	Dime 10	Quarter 25	Half Dollar 50
186	5	4	5	1	0
187	0	5	5	1	0
188	15	0	6	1	0
189	10	1	6	1	0
190	5	2	6	1	0
191	0	3	6	1	0
192	5	0	7	1	0
193	0	1	7	1	0
194	50	0	0	2	0
195	45	1	0	2	0
196	40	2	0	2	0
197	35	3	0	2	0
198	30	4	0	2	0
199	25	5	0	2	0
200	20	6	0	2	0
201	15	7	0	2	0
202	10	8	0	2	0
203	5	9	0	2	0
204	0	10	0	2	0
205	40	0	1	2	0
206	35	1	1	2	0
207	30	2	1	2	0
208	25	3	1	2	0
209	20	4	1	2	0
210	15	5	1	2	0
211	10	6	1	2	0
212	5	7	1	2	0
213	0	8	1	2	0
214	30	0	2	2	0
215	25	1	2	2	0
216	20	2	2	2	0
217	15	3	2	2	0
218	10	4	2	2	0
219	5	5	2	2	0
220	0	6	2	2	0
221	20	0	3	2	0
222	15	1	3	2	0
223	10	2	3	2	0

N	Penny 1	Nickel 5	Dime 10	Quarter 25	Half Dollar 50
224	5	3	3	2	0
225	0	4	3	2	0
226	10	0	4	2	0
227	5	1	4	2	0
228	0	2	4	2	0
229	0	0	5	2	0
230	25	0	0	3	0
231	20	1	0	3	0
232	15	2	0	3	0
233	10	3	0	3	0
234	5	4	0	3	0
235	0	5	0	3	0
236	15	0	1	3	0
237	10	1	1	3	0
238	5	2	1	3	0
239	0	3	1	3	0
240	5	0	2	3	0
241	0	1	2	3	0
242	0	0	0	4	0
243	50	0	0	0	1
244	45	1	0	0	1
245	40	2	0	0	1
246	35	3	0	0	1
247	30	4	0	0	1
248	25	5	0	0	1
249	20	6	0	0	1
250	15	7	0	0	1
251	10	8	0	0	1
252	5	9	0	0	1
253	0	10	0	0	1
254	40	0	1	0	1
255	35	1	1	0	1
256	30	2	1	0	1
257	25	3	1	0	1
258	20	4	1	0	1
259	15	5	1	0	1
260	10	6	1	0	1
261	5	7	1	0	1

	Penny	Nickel	Dime	Quarter	Half Dollar
N	1	5	10	25	50
262	0	8	1	0	1
263	30	0	2	0	1
264	25	1	2	0	1
265	20	2	2	0	1
266	15	3	2	0	1
267	10	4	2	0	1
268	5	5	2	0	1
269	0	6	2	0	1
270	20	0	3	0	1
271	15	1	3	0	1
272	10	2	3	0	1
273	5	3	3	0	1
274	0	4	3	0	1
275	10	0	4	0	1
276	5	1	4	0	1
277	0	2	4	0	1
278	0	0	5	0	1
279	25	0	0	1	1
280	20	1	0	1	1
281	15	2	0	1	1
282	10	3	0	1	1
283	5	4	0	1	1
284	0	5	0	1	1
285	15	0	1	1	1
286	10	1	1	1	1
287	5	2	1	1	1
288	0	3	1	1	1
289	5	0	2	1	1
290	0	1	2	1	1
291	0	0	0	2	1
292	0	0	0	0	2

Super Magic Squares

The following is a general formula for constructing a Supermagic 5 by 5 square.

What makes it Supermagic?

If you move the top row down to become the bottom row, it is still a magic square. Repeat this move 4 more times. Each time, the magic remains.

If you move the left column to become the rightmost column, it is still a magic square. Repeat this move 4 more times. Each time, the magic remains.

Let A, B, C, D, F, G, H, M, be any positive integers.

Set
Row 1, Column 1 = 2*A + B + C + D + F + G + H + M
Row 1, Column 2 = A + 2*B + C + D + F + G + H + M
Row 1, Column 3 = A + B + 2 * C + D + F + G + H + M
Row 1, Column 4 = A + B + C + 2*D + F + G + H + M
Row 1, Column 5 = F + G + H + M
Row 2, Column 1 = A + B + C + D + 2*F + G + H + M
Row 2, Column 2 = A + B + C + D + F + 2*G + H + M
Row 2, Column 3 = A + B + C + D + F + G + 2*H + M
Row 2, Column 4 = A + B + D + F + M
Row 2, Column 5 = A + B + 2*C + D + G + H + M
Row 3, Column 1 = B + D + F + H + M
Row 3, Column 2 = A + C + D + M
Row 3, Column 3 = A + B + C + D + F + G + H + M
Row 3, Column 4 = A+B+C+2*F+2*G+H + M
Row 3, Column 5 = 2* A + 2 * B + 2 * C + 2 * D + F + 2*G + 2*H + M
Row 4, Column 1 = A + B + C + F + 2*G + H + M
Row 4, Column 2 = 2*A + 2*B + 2*C + D + 2*F + 2*G + 2*H + M
Row 4, Column 3 = A + 2*B + C + 2*D + F + G + H + M
Row 4, Column 4 = C + D + H + M
Row 4, Column 5 = A + D + F + M
Row 5, Column 1 = A + B + 2*C + 2*D + G + H + M
Row 5, Column 2 = D + F + H + M
Row 5, Column 3 = A + F + G + M
Row 5, Column 4 = 2*A + 2*B + C + D + 2*F + G + H + M
Row 5, Column 5 = A + 2*B + C + D + 2*F + G + H + M

Example: Confirm that all of the below are magic squares with magic sum 2390.

573	486	518	556	257
570	576	545	273	426
245	213	478	590	864
498	878	564	185	265
504	237	285	786	578

Move top row to make new bottom row.

570	576	545	273	426
245	213	478	590	864
498	878	564	185	265
504	237	285	786	578
573	486	518	556	257

Again, move top row to make new bottom row.

245	213	478	590	864
498	878	564	185	265
504	237	285	786	578
573	486	518	556	257
570	576	545	273	426

Again, move top row to make new bottom row.

498	878	564	185	265
504	237	285	786	578
573	486	518	556	257
570	576	545	273	426
245	213	478	590	864

Again, move top row to make new bottom row.

504	237	285	786	578
573	486	518	556	257
570	576	545	273	426
245	213	478	590	864
498	878	564	185	265

Move leftmost column to rightmost position.

486	518	556	257	573
576	545	273	426	570
213	478	590	864	245
878	564	185	265	498
237	285	786	578	504

Again, move leftmost column to rightmost position.

518	556	257	573	486
545	273	426	570	576
478	590	864	245	213
564	185	265	498	878
285	786	578	504	237

Again, move leftmost column to rightmost position.

556	257	573	486	518
273	426	570	576	545
590	864	245	213	478
185	265	498	878	564
786	578	504	237	285

Again, move leftmost column to rightmost position.

257	573	486	518	556
426	570	576	545	273
864	245	213	478	590
265	498	878	564	185
578	504	237	285	786

Geometric Series

The teacher called it an opportunity test! Diane half smiled as she tilted her head back and using her fingers, brushed her long black hair over and behind her ears. Mr. Smiley called it an opportunity test because it gave everyone a chance to make an A for the year. It was Mr. Smiley's tradition. Anyone who made an A on his math opportunity test on the last day of school would earn an A for the year.

Diane intended to make that A. Today she had Explorer with her. Explorer was Diane's magical friend that had helped her for as long as she could remember. They could talk silently to each other. Explorer would make sure she made a perfect grade.

Diane looked at the first problem. Add 1024 + 2048 + 4096 + 8192 + 16384. Diane asked Explorer "Well, do you know the answer?"

Explorer sounded amused. "Your teacher is making it easy on himself. This is a doubling series. Look at the first two numbers, 1024 and 2048. What is 1 + 2?"

Diane looked again. "1 + 2 is 3. So what?"

Explorer said "You will see. The third number is 4096. What is 3 + 4?"

Diane said "7. So 1 + 2 + 4 is 7. And 7 is 1 less than 8."

Explorer said "Now you are getting it.

Diane looked at the last two numbers 8192 and 16384. "7 + 8 is 15 and 15 is one less than 16. I see that 1 + 2 + 4 + 8 + 16 is 31 because 31 is 1 less than twice 16."

Explorer said "You have it right. How can you be sure of it?"

Diane had a sudden inspiration. "Suppose it had been 1 + 1 + 2 + 4 + 8 + 16. Since 1 + 1 is 2, it would have been equal to 2 + 2 + 4 + 8 + 16. But since 2 + 2 is 4, it would have been equal to 4 + 4 + 8 + 16. But since 4 + 4 is 8, this would be equal to 8 + 8 + 16. And since 8 + 8 is 16 this would be equal to 16 + 16 = 32. So, If 1 + 1 + 2 + 4 + 8 + 16 is 32, then 1 + 2 + 4 + 8 + 16 must be 1 less than 32 which is 31.

Explorer said "Now you are ready to do the original problem. What is 1024 + 2048 + 4096 + 8192 + 16384?"

Diane saw that 1024 + 1024 was 2048. And 2048 + 2048 was 4096. And 4096 was halfof 8192. And 8192 was half of 16384. She said "I see that this really is a doubling series. If it were 1024 + 1024 +2048 + 4096 + 8192 + 16384 then the answer would be twice 16384. Twice 16384 is equal to 32768. This means that 32768 is 1024 too much. So the correct sum is 32768 - 1024 which is equal to 31744." Diane wrote down her answer.

Diane looked at the second question. Add 21 + 42 + 84 + 168 + 336 + 672. "aha", Diane thought, "this is another doubling series." Diane verified that 21 was half of 42, that 42 was half of 84, that 84 was half of 168, that 168 was half of 336 and that 336 was half of 672. Diane thought, "If this had been 21 + 21 + 42 + 84 + 168 + 336 + 672 then the answer would have been twice 672 which is equal to 1344. Therefore the correct answer is 1344 - 21. Diane wrote down 1323 for her answer to the second question.

Diane looked at the third question. Add 3 + 6 + 12 + 24 + 48 + 96 + 192 + 384. Diane verified that this was a doubling series as she read the numbers. She then doubled 384 and subtracted 3. Diane smiled as she wrote 765 down for the answer to the third question. "This is fun", she said.

The fourth question was different:512 - 256 + 128 - 64 + 32 - 16 + 8 - 4 + 2 - 1 = ?

Explorer waited. "You do notice that each number is twice the value of the following one?"

"So what?" Diane tossed off casually, then she looked again. "Wait, now I see... 256 from 512 is, well, half, or 256. And for 64 from 128, it's again half, and so on. This isn't a doubling series, it's a 4-times series! You can split it into pairs, and each pair is really just half of the first number." She quickly rewrote the problem: (512-256) + (128-64) + (32-16) + (8-4) + (2-1) =256 + 64 + 16 + 4 + 1 = ? Having found the pattern, Diane was only momentarily stumped. "How would I add it, hmm."

"How many more..." Explorer began, only to be quickly interrupted.

"You think too fast," chided Diane. "To make the 4, I'd need three more 1's, but then I still need two more 4's to make the 16, so I really need two more of everything."

Explorer said "Very good Diane. So what is the answer?"

Diane said, "When I subtract the 1 from 1024 to get 1023 I still have every number 3 times. So I have to divide by 3 to get the correct answer." Diane divided 1023 by 3. Diane smiled as she wrote 341 for the answer to the fourth question.

Diane looked at the fifth question. Add 729 + 364 + 182 + 91 + 45 + 22 + 11 + 5 + 2 + 1. "Help me Explorer. This is not a doubling series. What trick is Mr Smiley playing on us now?"

Explorer said "This is a halving series. It is almost a doubling series in reverse. What would you have to add to the 2 + 1 to make 2 + 2?"

Diane said, "I would have to add 1 of course. However, 2 + 2 makes only 4, not 5. I'm stuck."

Explorer said "Not really. Just note that you have to add another 1 to the 4 to make 5. Then you can add 5 + 5."

Diane said "But how does that help? 5 + 5 is 10, not 11. Wait. Now I get it. I just count the times I have to add 1 to get to the next number. If the next number is odd then I have to add 1 to get to it. If the next number is even, I am already to it. So the number of times I have to add 1 is the number of odd numbers in the halving series."

Diane counted the number of odd numbers in the halving series 729 + 364 + 182 + 91 + 45 + 22 + 11 + 5 + 2 + 1. "There are 6 odd numbers in the halving series. So the answer is 6 less than twice 729." Diane doubled 729 to get 1458, and then subtracted 6. Diane wrote down 1452 for the answer to the fifth question.

Explorer said "You have the answer right Diane, but only because the last number in the halving series was 1. If it had been different than 1, you would have gotten a wrong answer by your rule."

"So what rule should I have used?" asked Diane.

Explorer said. "The rule should have been that the answer is twice the first number minus the last number minus the number of odd numbers not counting the last number."

Diane said, "I see. Twice 729 is 1458. The last number is 1. The number of odd numbers not counting the last number is 5. So the answer is 1458 - 1 - 5. I get the same answer 1452."

In this way, Explorer helped Diane earn an A in her seventh grade math class for the year.

You too can explore numbers. Make up your own opportunity test and try to make an A on it.

N to the fourth power plus 4

Can you factor $n^4 + 4$

Method 1

If n = 1, then $n^4 + 4 = 1 + 4 = 5 = 1 * 5 = (-1)*(-5)$
if n = 2, then $n^4 + 4 = 16 + 4 = 20 = 2 * 10 = 4 * 5 = (-2)*(-10) = (-4) * (-5)$
if n = 3, then $n^4 + 4 = 81 + 4 = 5 * 17 = (-5) * (-17)$

If $n^4 + 4$ factors into two quadratic polynomials, p1 and p2,
then we could require that

p1(1) = 1
p1(2) = 2
p1(3) = 5

p2(1) = 5
p2(2) = 10
p2(3) = 17

$p1(n) = a1\ n^2 + b1\ n + c1$

9 a1 + 3 b1 + c1 = 5
4 a1 + 2 b1 + c1 = 2
  a1 +   b1 + c1 = 1

Subtract each equation from one above it.

5 a1 + b1 = 3
3 a1 + b1 = 1

Subtract each equation from one above it.

2 a1 = 2

a1 = 1
b1 = -2
c1 = 2

p1(n) = n^2 - 2 n + 2

Set
p2(n) = a2 n^2 + b2 n + c2.

p2(3) = 17
p2(2) = 10
p2(1) = 5

9 a2 + 3 b2 + c2 = 17
4 a2 + 2 b2 + c2 = 10
 a2 +   b2 + c2 = 5

Subtract each equation from one above it.

5 a2 + b2 = 7
3 a2 + b2 = 5

Subtract each equation from one above it.

2 a2 = 2

a2 = 1
b2 = 2
c2 = 2

p2(n) = n^2 + 2 n + 2

Confirm

(n^2 - 2 n + 2) (n^2 + 2 n + 2) = n^4 + 4

Method 2

Factor $N^4 + 4$

Test if the factorization is of the form

$(a_1 N^2 + b_1 N + c_1)(a_2 N^2 + b_2 N + c_2)$.

Multiply to get

$a_1 a_2 N^4 + (a_1 b_2 + b_1 a_2) N^3 + (b_1 b_2 + a_1 c_2 + c_1 a_2) N^2 + (b_1 c_2 + c_1 b_2) N + c_1 c_2 = N^4 + 4$

Equating coefficients, we get

$a_1 a_2 = 1$
$(a_1 b_2 + b_1 a_2) = 0$
$(b_1 b_2 + a_1 c_2 + c_1 a_2) = 0$
$(b_1 c_2 + c_1 b_2) = 0$
$c_1 c_2 = 4$

From $a_1 a_2 = 1$, we conclude that $a_2 = a_1$.

a_2 might be 1 or -1.

We first try $a_1 = 1$; $a_2 = 1$.

From $c_1 c_2 = 4$, it might be that $c_1 = 1$, $c_2 = 4$, or $c_1 = c_2 = 2$.

We first try $c_1 = c_2 = 2$.

Equations are reduced to

$b_2 + b_1 = 0$
$b_1 b_2 + 4 = 0$
$2 (b_1 + b_2) = 0$

Thus we see that $b_1 = -b_2$
and $b_2^2 = 4$

We first try $b_2 = 2$, $b_1 = -2$.

Confirm

$(N^2 - 2 N + 2)(N^2 + 2 N + 2) = 0$

Method 3

Factor $N^4 + 4$

Add and subtract $(4 N^2)$ to make the difference of two squares.

$N^4 + 4 = N^4 + 4 N^2 + 4 - 4 N^2 = (N^2 + 2)^2 - (2 N)^2$
$= (N^2 - 2 N + 2) (N^2 + 2 N + 2)$

www.ingramcontent.com/pod-product-compliance
Lightning Source LLC
Chambersburg PA
CBHW081303170526
45165CB00011B/3392